Pro/ENGINEER Solutions™ and Plastic Design

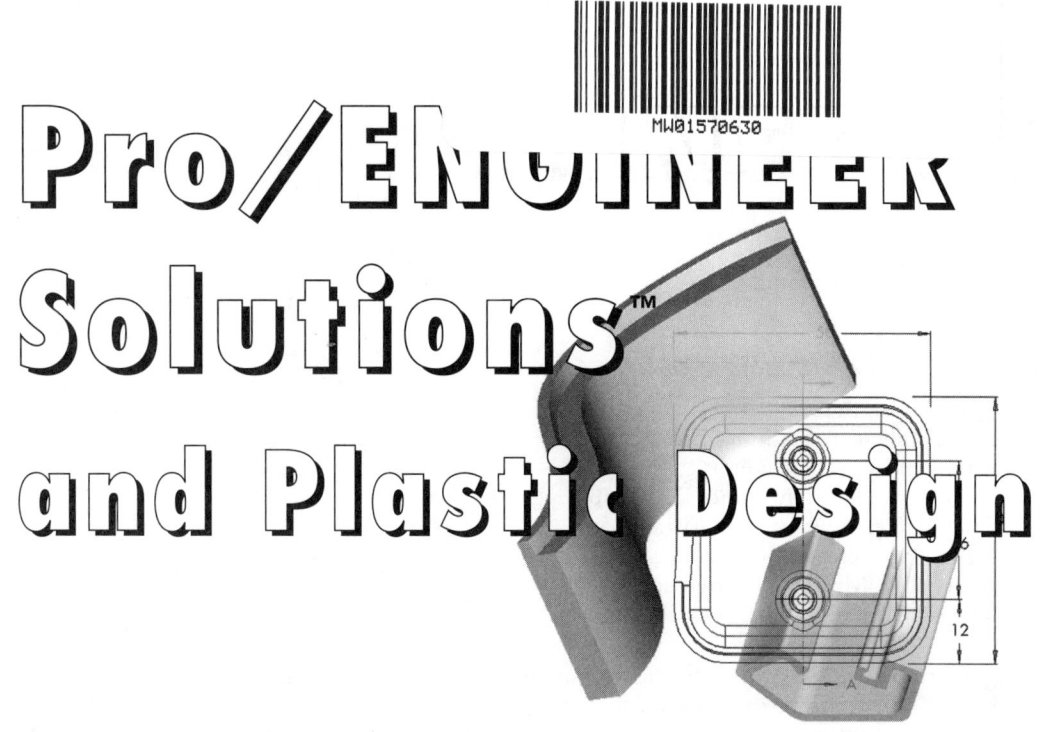

Norman Ladouceur, CET
John McKeen, Ph.D.

Pro/ENGINEER® Solutions™ and Plastic Design
By Norman Ladouceur and John McKeen

Published by:

OnWord Press
2530 Camino Entrada
Santa Fe, NM 87505-4835 USA

Carol Leyba, Publisher
David Talbott, Acquisitions Director
Daril Bentley, Senior Editor
Andy Lowenthal, Director of Production and Manufacturing
Cynthia Welch, Production Manager
Liz Bennie, Director of Marketing
Deborah R. Mrantz, Cover Design
John McManaman, Indexer

All rights reserved. No part of this book may be reproduced or transmitted in any form or by any means, electronic or mechanical, including photocopying, recording, or by any information storage and retrieval system, without written permission from the publisher, except for the inclusion of brief quotations in a review.

Copyright © 1999 OnWord Press

SAN 684-0269

10 9 8 7 6 5 4 3 2 1

Printed in the United States of America

Library of Congress Cataloging-in-Publication Data

Ladouceur, Norman, 1956–
Pro/ENGINEER Solutions and Plastic Design / Norman Ladouceur and John McKeen.
 p. cm.
Includes index.
ISBN 1-56690-188-X
1. Pro/ENGINEER. 2. Engineering design—Data processing.
3. Plastic analysis (Engineering)—Data processing. 4. Plastic design—Data Processing.
5. Computer-aided design.
I. McKeen, John, 1953–   . II. Title.
TA174.L331999
668.4'0285'569—dc2198-54937
CIP

Trademarks

OnWord Press is a registered trademark of High Mountain Press, Inc. Pro/ENGINEER is a registered trademark of Parametric Technology Corporation. Pro/SURFACE is a trademark of Parametric Technology Corporation. Other products and services mentioned in this book are either trademarks or registered trademarks of their respective companies. OnWord Press and the authors make no claim to these marks.

Warning and Disclaimer

This book is designed to provide information about Pro/ENGINEER as it relates to plastics design. Every effort has been make to make this book complete and as accurate as possible; however, no warranty or fitness is implied.

The information is provided on an "as is" basis. The authors, Parametric Technology Corporation, and OnWord Press shall have neither liability nor responsibility to any person or entity with respect to any loss or damages in connection with or arising from the information contained in this book.

About the Authors

Norman Ladouceur is a mechanical technician with the Corporate Design Group at Nortel Networks of Ottawa, Canada. Norm is also a certified engineering technician in the province of Ontario, Canada. His CAD background dates back to 1981, when he was first introduced to Computervision software. His computer software experience includes CADAM, Graftek, CATIA, and Pro/ENGINEER since version 5. His product development background includes residential, personal, and business telephone terminals, power tools, and household appliances. Norm is the author of *INSIDE Pro/SURFACE*, also published by OnWord Press.

John McKeen is a mechanical engineer at Nortel Networks of Ottawa, Canada. John has a Ph.D. in mechanical engineering and is a member of the Professional Engineers of Ontario. He has been involved with computer-aided design (CAD) and computer-aided engineering (CAE) since 1982. His computer software experience includes Graftek, SDRC, CATIA, MARC, PATRAN, and Pro/ENGINEER since version 2. He introduced one of the first rapid prototyping systems, the 3D Systems Stereolithography Apparatus (SLA) equipment, into Nortel in 1989. He presented the first paper, for the Rapid Prototyping Society in 1994, that showed how to make SLA molds for plastic injection molding machines. He has performed design and analysis for residential and business telephone terminals, and for small to medium sized communications equipment.

Acknowledgments

We would like to thank the management and staff at Nortel Networks who supported us in this project and helped provide the valuable plastic design experience. Thanks also to our families. Without their encouragement, this project would not have been realized. Special thanks to Paul McDonald for taking the time to proofread the initial manuscript. His patience, questions, and suggestions have made the book more technically complete. Thanks also to Mike Dolson, Dennis Steffen, and Tony Langenberg for their technical reviews. Last, but not least, thanks to Daril Bentley and the staff at HMP for all of their help during the past few months.

Contents

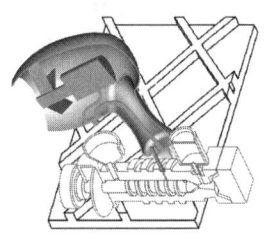

Introduction   xiii

    Purpose, Content, and Audience xiii
    Book Structure and Features xv
        Structure xv
        Features xv

Part I   Fundamentals   1

Chapter 1   Plastic Molding and Fabrication   3

Introduction 3
Plastic Tooling Descriptions and Terminology 3
    The Mold 4
    The Core 5
    The Cavity 5
    The Parting Line 6
    Shrink 7
    Draft 7
    Ejector Pins 9
    Slide 10
    Sprue, Gates, and Runners 12
    Cooling Line 13
Electrical Discharge Machining 14
    Electrodes 14
    Vents 15
Plastic Part Fabrication Methods 15
    Traditional and Thin Wall Injection Molding 16
        The Process 16
        Cost Considerations 20
    Gas Assisted Injection Molding 21
    Foamed Parts 23

>   Extrusion Molding 24
>   Blow Molding 26
>   Rapid Prototyping and Tooling Methods 30
>   Rapid Prototyping 30
>   Rapid Tooling 31
>   Epoxy Molds 31
>   Plastic Molds 32
>   Summary 33

Chapter 2   Why Select Plastics?   35

>   Introduction 35
>   Electrical Insulation Properties 36
>   Cosmetic Appearance 39
>   Versatile Part Design of Complex Parts 41
>   Flexible Material 42
>   Plastics Properties 43
>   Snap Feature 45
>   Other Material Property and Design Considerations 45
>   Energy Absorption 45
>   Deflection Resistance 48
>   Durable Surface Finish 49
>   Functioning Part Design 49
>   Associated Costs 50
>   Matching Design and Plastics Technology 51
>   Strong, Hard, and Tough Materials 51
>   Deflection 52
>   Temperature 52
>   Other Physical Properties 53
>   Summary 53

Chapter 3   Design Philosophy   55

>   Introduction 55
>   Initial Design Planning 56
>   Considering the End User 56
>   Technology Limitations 58
>   Visualizing Product Designs 59
>   Top-down Design Approach 60

  Identifiable Restrictions 60
  Bottom-up Design Approach 64
  Video Game Controller Wrap-up 67
 Identifying Components in Product Design .67
  Splitting a Design into Top and Bottom Components 69
  Splitting Components into Front and Back Piece Parts 70
  Splitting Components into Left and Right Piece Parts 71
  Electric Razor Wrap-up 72
 Product Design Modeling Techniques 73
  Curling Iron Skeleton Model 74
  Product Individual Piece Part Creation .76
  Product Design Using a Master Part File 77
 Summary 81

Chapter 4 Design Planning and Organization 83

 Introduction 83
 Parting Surfaces and Parting Planes 84
 Identifying Root Features 87
 Identification of Critical Specifications 92
 Logical Feature Creation Process 95
  Logical Feature Creation Example: A Housing 96
  Tips for Identifying Buried Features 98
  Creating a Model Plan 100
 Component Fastening 101
  Ultrasonic Welding 102
  Snap Fit Joints 104
   Testing Snap Fit 105
   Modeling "Steel Safe" 106
  Using Screw Bosses 109
  Wall Mount Holding Tabs 112
 Summary 112

Part II The Tool Box 113

Chapter 5 Plastics Design Features 115

 Introduction 115
 Shells of Constant Wall Thickness 115

Shells of Irregular Wall Thickness 117
Manually Creating a Shell 118
Draft 120
    No Split 121
    Split at a Curve or Quilt 121
    Using Split as a Sketch 122
Features Aiding Plastics Design 123
    Rounds 123
        Simple Rounds 124
        Advanced Rounds 129
    Crowning 130
Local Thinning and Thickening of Material 131
    Ears 131
    Lip Feature 132
    Replacing a Surface 133
    Reminders About Mold Creation 133
Summary 134

Chapter 6   Strength of Shapes   135

Introduction 135
Component Shape and Overall Part Strength 135
    Solving Heat Dissipation Problems 135
    Guidelines for Bending and Deflection 139
    Quick Impact or Shock Guidelines 141
Rib and Web Design Techniques and Rules 143
    Straight Ribs 143
    Rotational Ribs 144
    Techniques and Rules 145
    Ribs and Draft 146
    Curved Ribs and Draft 148
    Ribs and Thin Wall Designs 148
    Other Applications for Ribs 149
Summary 149

Chapter 7   Designing for Manufacturability   151

Introduction 151

Modeling Techniques That Enhance Material Flow 151
Managing Sink Marks, Flash, and Weld Lines 156
  Sink Marks 156
  Managing Flash and Weld Lines 158
Inserted Areas 159
Managing Tolerances and Relations 160
  Maximum/Minimum Tolerance Modeling 160
  Dimension Relations 161
Evaluating Features 163
Summary 165

Part III   Applications   167

Chapter 8   Portable Compact Disc Player   169

Introduction 169
Project Description 170
Applying Design Philosophy 171
Modeling Plan and Process 173
  Intelligent Start Part Plan 175
  Create Product External Geometry 176
  Conclusions: Intelligent Start Part 197
  Master Part Approach to Modeling the CD player 197
    Design Philosophy and Master Part Plan 198
    Model the Components 200
    Conclusions: Master Part Technique 201
  Skeleton Model Technique 201
Summary 202

Chapter 9   Replacing Several Components with One Plastic Part   203

Introduction 203
Project Description 204
Design Planning 205
  Modeling Plan 206
  Modeling Process 207
Summary 215

Chapter 10   Component Fastening   217

Introduction 217
Holding Components in Desired Locations 218
Heat Staking 218
    Staking Pins 219
    Crush Ribs 219
Ultrasonic Welding 221
    Understanding the Process 221
    Butt Weld Joint Method 223
    Energy Director Method 225
Snap Fit Design 227
    Modeling Process 228
    Finished Snap 233
    Screws 234
Summary 236

Chapter 11   Thin Wall Component Design   237

Introduction 237
What Is Thin Wall Design? 237
Design Philosophy 238
Modeling Plan and Process 239
    Modeling Plan 239
    Modeling Process 240
Summary 249

Chapter 12   Extruding Shapes   251

Introduction 251
Avoiding Warpage 252
Plastics for Support Structures 253
Post Processing on Plastic Extrusions 254
Summary 255

Chapter 13   Blow Mold Designs   257

Introduction 257
Designing a Bottle for Blow Molding 258
    Neck Section 258

Contents

    The Basic Shape 260
    Top Section 260
    Labeling 261
    Bottom 262
    Shell and Volume 263
Summary 263

Chapter 14 Mirror Parts 265

Introduction 265
A Mirroring Project 265
    Project Description 266
    Latch Modeling Process 266
Summary 269

Chapter 15 Volume Calculations 271

Introduction 271
Calculating Volumes 271
    Volume of Plastic in a Part 272
    Volume Within a Plastic Part 272
Summary 276

Chapter 16 Applying Material and Mass Properties 277

Introduction 277
The Material File 277
    Material File Terms 278
    Material Files and Finite Element Analysis 280
The Mass Properties File 281
    Center of Gravity 281
    Moment of Inertia 281
    Principle Moments of Inertia 282
    Radii of Gyration 282
    Cross Section Mass Properties 283
Summary 285

Chapter 17 Information and Clearance/Interference Tools 287

Introduction 287

    Numerically Based Information Tools 288
        Info Measure 289
            Using Datum Points 289
            Edges and Curves 290
            Minimum Radius 291
            Thickness 291
            Draft Check 292
        Visual Analysis Tools 294
            Gaussian Curvature 295
            Slope Analysis 295
            Porcupine Analysis Tool 296
            Surface Normal Vectors 299
            Shading 300
        Clearance/Interference Checking of Assemblies 302
            Clearance Checking. 302
            Interference Checking 303
    Making Analysis Tool Use a Habit 307
    Summary 307

Chapter 18 Imported Geometry 309

    Introduction 309
    Geometry Source 310
    What You Can Import 310
    Defining Import Requirements 312
    Features Difficult to Import 313
    Redefine Tools for Imported Geometry 314
    IGES and STEP 315
        IGES – Initial Graphics Exchange Specification 315
        STEP – Standard for the Exchange of Product Model Data 315
        Import Geometry Project 316
        LOG Files 321
        Dealing with Change in Imported Geometry 321
            Re-import Example 322
            Logical Import Process 325
    Summary 327

Index 329

Introduction

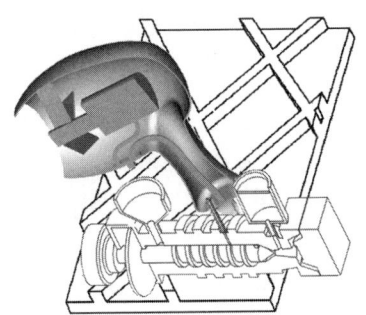

Pro/ENGINEER is widely used as a design tool for plastic components in the computer/consumer product categories. As Pro/ENGINEER expands into the automotive and industrial categories, there will be an increase in the use of the tool to develop plastic component/product designs. Plastics technology is changing rapidly to conform to new demands such as thin wall molding to reduce material consumption and to address environmental product design concerns. Readers of this book are encouraged to seek out the appropriate technology for their designs through consultation with their plastic component or resin suppliers.

Purpose, Content, and Audience

The goal of this book is to help the Pro/ENGINEER user gain the knowledge and know-how that allow the designer to create robust, modifiable geometry in a manner directly portable to plastics tooling suppliers. Consideration is given to subject areas relating to the mass production of plastic components. Pro/ENGINEER users of all levels will find discussions of interest in this book.

The material in this book is closely related to other pouring and injection technologies such as sand casting with iron and aluminum and die-casting of aluminum and magnesium. Reference is occasionally made to these other technologies where it is appropriate to the discussion.

In addition to a general audience, this book is intended to target an industry-specific market for plastic manufacture of components and products. The design information and Pro/ENGINEER modeling techniques are intended to suit that market. Throughout the book, terminology is used that is common to the plastics industry. This terminology is explained in the early sections of the book, and is referred to throughout.

Have you ever wondered why draft plays such a role in plastic tooling and design? Why is the draft used on a product exterior 2 degrees or more? These questions and more are answered as *Pro/ENGINEER Solutions and Plastic Design* takes you through the design process from concept to manufacturable product.

The applications section of the book focuses on using the knowledge found in the first sections of the book and applying that knowledge to your design. This book is not intended to eliminate the need for your PTC user manuals but adds to that information in terms of focusing on the how, when, and why of the design job specific to the plastics industry.

It is no accident that a plastic product looks and feels like a quality piece of merchandise. This is a result of good design, down to the component level. The factors that relate directly to the quality of the product can start with the material and technology chosen for the task and work all the way down to the location and placement of the internal rib design for individual components. But what happens when you finish a design? How do you know if it's manufacturable? Questions such as these are answered in the information and analysis sections of the book as it identifies and explains the tools Pro/ENGINEER offers for creating and evaluating plastics designs.

Book Structure and Features

Structure

Pro/ENGINEER Solutions and Plastic Design is divided into four parts: Fundamentals, The Tool Box, Applications, and Component Analysis and Mass Property Calculations.

- Part I, Fundamentals, includes chapters on molding and fabrication, reasons for selecting plastic as the material of choice, design philosophy, and design planning and organization.

- Part II, The Tool Box, covers design features, shape strength and related issues, and manufacturable design.

- Part III, Applications, takes you through discussions that expand on and refine previous topics via hands-on projects, including the design of a compact disc player, the replacement of several components with a single part, and other projects offered within each chapter. Other discussion and project topics include component fastening, thin wall component design, extrusions, design for blow molding, and mirror image part creation.

- Part IV, Component Analysis and Mass Property Calculations, explores volume calculations, the application of material and mass properties, Pro/ENGINEER's information and clearance/interference tools, and dealing with imported geometry.

Features

Introductions and Summaries

Chapters within this book begin with an introduction and end with a summary. These sections inform you of what to expect to learn from a chapter and the highlights of what you should take away with you after having completed a chapter.

Illustrations

Illustrations support text discussion throughout, offering visual information where verbal explanation becomes complicated. Illustrations also depict stages of the hands-on project exercises found in Part III.

Bulleted and Numbered Lists

The requirements of a process, the significant features of a design concept, and similar items are generally set off as bulleted lists. Numbered lists designate the steps within an example process for purposes of discussion, or the steps (where applicable) within the "phases" of a hands-on project exercise.

NOTEs and TIPs

NOTEs and TIPs, such as the following, inform you of points to keep in mind (NOTEs) and efficient methods of dealing with a problem or performing a task (TIPs).

Project Exercises

Hands-on project exercises—found mainly in Part III, Applications, but elsewhere in support of text discussion—are easily located, indicated by a bar that runs down the left-hand margin. Project exercises are divided into Phases, which represent the stages of the design process. Where appropriate, Phases are divided into numbered steps.

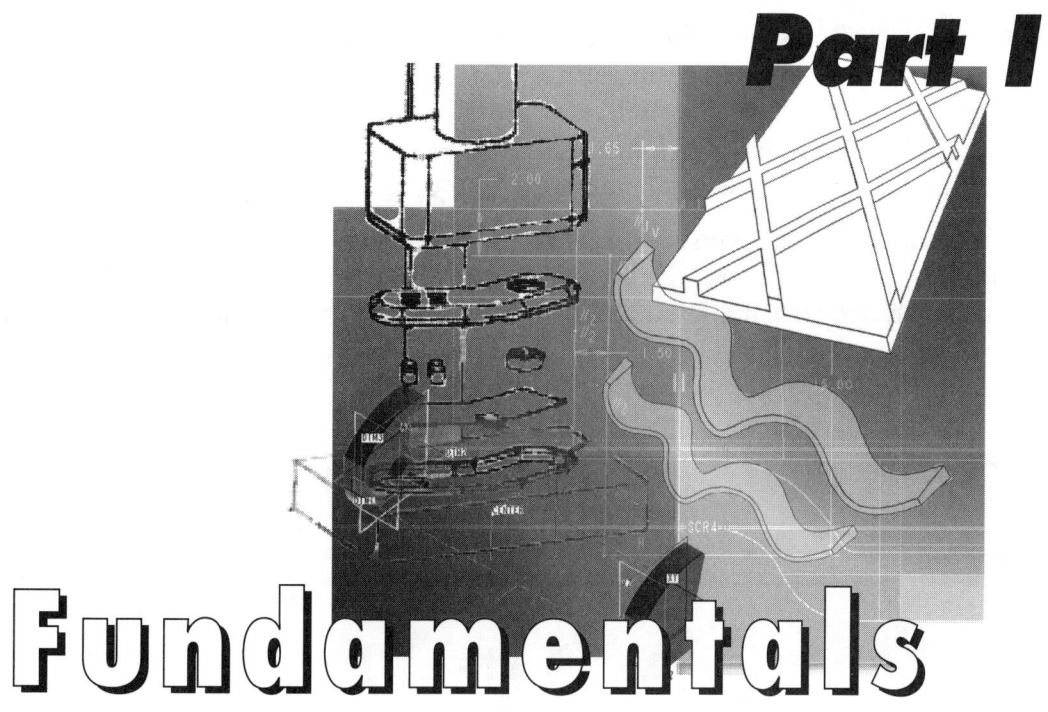

Part I

Fundamentals

Pro/ENGINEER provides users with the tools necessary to design plastic components with a high level of confidence in the overall quality of their design. Pro/ENGINEER is readily adaptable to various plastics molding technologies. Part I describes the various methods of plastics molding technology and how Pro/ENGINEER fits into this picture, assisting you in designing components. Part I also explores the versatility of Pro/ENGINEER as your tool of choice.

Molding terms and fabrication methods are briefly explained to provide you with an understanding of the "language of plastics," including reference sources for especially confusing terminology. These terms are used in the context of Pro/ENGINEER as the modeling tool.

Procedures are presented for the design of plastic parts using the various fabrication methods. Modeling similarities and differences are described for each fabrication method, which will aid you in developing the ability to make appropriate choices in model creation using Pro/ENGINEER. The reasons for selecting plastic as the design material are also explored.

The discussion also covers the areas of design philosophy and planning, which will help you develop the ability to evaluate design procedures and "plans of attack." By understanding modeling issues early in the design cycle, as this book attempts to help you do, you will save valuable and in some cases significant time in the overall product design cycle.

Chapter 1
Plastic Molding and Fabrication

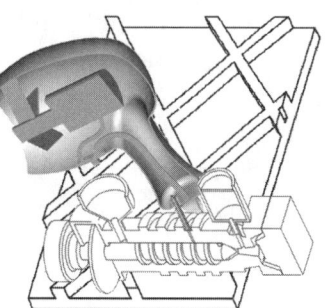

Introduction

Plastic components can be fabricated by a number of methods. Each method uses a different technology to produce the final part. The technology selected is determined by the end use of the plastic part and by the physical means to be used to create the part. This chapter describes the most common fabrication methods used for plastic part manufacture. The main plastic design requirements for each fabrication method are described.

Plastic Tooling Descriptions and Terminology

The following sections define and describe the most common components of plastics manufacture. These sections cover terms (part of the "language of plastics") with which any designer of plastic parts should be familiar. These components and terms central to the plastic parts industry are expanded on throughout the rest of the book in terms of using Pro/ENGINEER to design and create plastic parts.

The Mold

Plastic parts are solid, as opposed to liquid or vapor, at room temperature. The plastic parts you use and encounter every day begin in viscous (liquid) form before being molded into the shape you see. In order for the viscous plastic to take its final form, it must be put into a volume that will create the desired shape. This volume is called a mold—a physical device that gives a plastic part its physical shape and volume. A cross section of a typical mold is shown in the following illustration.

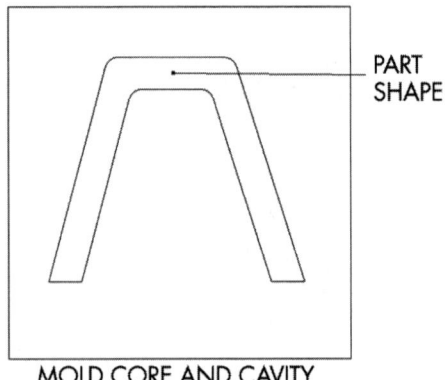

Cross section of a mold.

As previously stated, a mold is an enclosed volume, and it is assumed that the mold can hold viscous plastic without any of it leaking. The overall shape and volume of a final plastic part are determined by the mold used to create the part. A mold also defines what a part's contours and textures will be like in final form. A mold—which might be made of steel, aluminum, urethane, or plastic—also defines the wall thickness of a plastic part.

Given the often intricate and convoluted shapes of parts, it can take a long time to design and manufacture a mold. Because of this, and because of the expense of their creation, molds are commonly used for as many cycles of manufacture as possible. Sometimes millions of parts are made from duplicates of the initial mold. In terms of reuse, molds are designed in such a way that the viscous plastic can be brought into the mold and removed as a solid plastic part without

Plastic Tooling Descriptions and Terminology

having to break and destroy the mold. This implies that a mold must consist of parts that perform the required handling of viscous material and that can be separated for removal of the final product. A mold fundamentally consists of a core and a cavity. These parts are defined and described in the sections that follow.

The Core

As previously noted, once viscous plastic has been placed inside a mold and has solidified, there must be some method of removing the plastic from the mold. In this regard, a mold typically consists of two major parts so that the mold can be taken apart and the plastic part removed without destroying the mold.

One major part of the mold is the core. The core is that part of a mold that forms the "inside" shape of a part. For example, the plastic drinking cup in the following illustration exhibits both "inside" and "outside" surface and volume areas. The core of a mold shapes the inside contour of such an object—in this case, a cup.

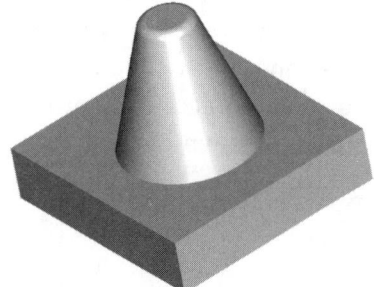

Mold core area of a plastic drinking cup.

The Cavity

The second major part of a mold is its cavity. The cavity is the part of the mold that forms the outside shape of a part. For example, the outside contour of a plastic drinking cup is shaped by the cavity of the mold, as shown in the following illustration.

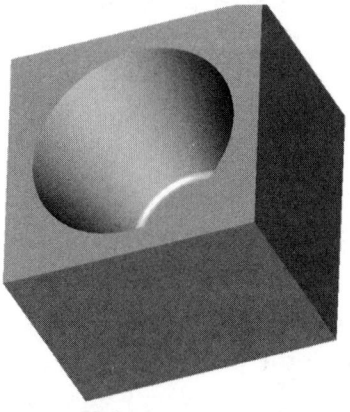

Mold cavity for a plastic drinking cup.

The Parting Line

A plastic part is obtained by removing the part from a mold's solid core and cavity (hollow). To remove the part, the core and cavity are typically pulled apart from each other on a plane. Core and cavity are separated at what is known as a parting line such that they move away from each other at 180 degrees at all places along this line. The parting line visible (if at all) on a part is created where the core and the cavity of a mold contact each other at the outer surface of the part. The parting line of a mold is shown in the following illustration.

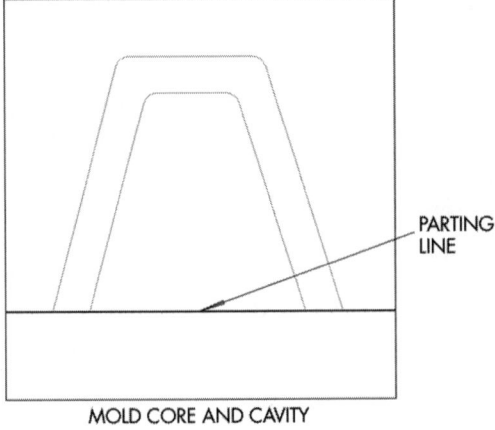

Parting line of a mold.

Plastic Tooling Descriptions and Terminology

Shrink

When a liquid or otherwise viscous plastic is placed in a mold, its temperature is high. A hot plastic material takes up more volume than the same material at, say, room temperature. When the plastic cools, its volume decreases, which is referred to as "shrink." After a hot plastic material is placed into a mold, filling the mold, it shrinks to a smaller volume as the plastic is cooled. Where the plastic shrinks, stresses are placed on the core and cavity of the mold. The following illustration shows a cross-sectional picture of a plastic cup, indicating where shrinkage would occur and where stresses would be placed on the core and cavity of the mold.

Shrink and stresses formed by cooled plastic in a mold.

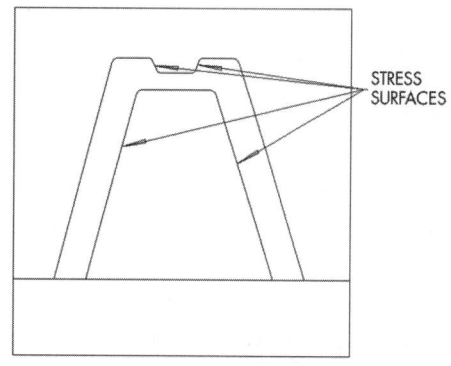

Draft

Once plastic has cooled sufficiently in a mold, the core and cavity may be pulled apart. If the plastic is still warm when core and cavity are pulled apart, the plastic may not stay in the desired final shape. This is like taking an ice cube out of an ice tray before the water has completely frozen. The ice cube is weak and its wall may break, letting the water flow out.

If the walls of the core and cavity are perpendicular to the parting line, and the core and cavity are pulled apart, stresses placed on the sides of the core and cavity due to the shrinkage of the plastic will cause the plastic part to break or deform at the stress surfaces. To remedy this situation, an angle (called the "draft angle") is put on the core and cavity where shrinkage and stresses are formed on the surfaces. This angle changes the stress forces, due to shrinkage, from being perpendicular to the core and cavity to being at an angle that aids in the removal of the plastic part. "Draft" is the angle from perpendicular to the parting line. For smooth walls, the part does not have as much shear stress. Therefore, the draft may be as small as 1/2 degree. The illustration that follows shows the draft angle for a plastic drinking cup.

Draft angle for a plastic drinking cup.

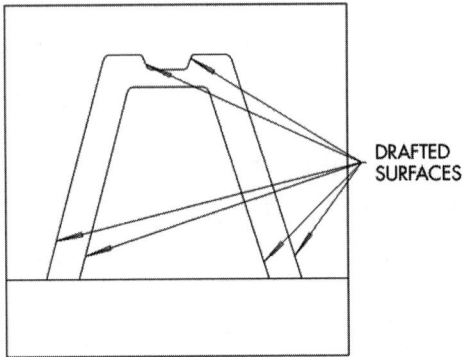

Surfaces are often textured to make them functional or cosmetically appealing. If a part has a textured or otherwise rough or bumpy surface, the minimum draft angle might be 5 degrees. The following illustration shows an exaggerated example of a textured surface for a plastic cup. It is textured in the middle to give a gripping surface to the cup.

Plastic Tooling Descriptions and Terminology

Textured surface on a drinking cup.

"Draft angle" in mold design is used in manufacturing processes other than plastic parts. A draft angle of 5 degrees or greater is required for molds consisting largely of sand because the rough surface causes high shear stress. Sand molds are used for metal part fabrication or metal castings. The texture of the sand forces the draft angle to be large or the part cannot be removed from the mold.

Ejector Pins

Often, when core and cavity are pulled apart, the plastic part does not fall off or out of the core of the mold. Devices are needed to assist in the removal of the plastic part from the core. These devices are called ejector pins, an example of which is shown in the following illustration. They are usually thin metal pins that push, from the core side (shown in the illustration), on the cooled plastic part. They make the plastic part fall from the molding machine to a collection tray. The ends of an ejector pin are curved, as necessary, to follow the shape of the core surface.

Ejector pin.

Ejector pins are usually found on the core side of a mold. However, it is sometimes necessary that ejector pins be placed on the cavity of a mold because the part cannot be removed without them. Ejector pins leave flash lines that mar the surface of the part and do not leave the surface looking aesthetically pleasing. The mark left is the joint between the ejector pin and the core.

Because the ejector pin cannot fit exactly into the core (otherwise, it would not be able to move and push the part away from the core), there must be a finite space between the core and the pin. This space forms the flash line. Mold designers try to place ejector pins where flash lines left behind will not be seen. Fortunately, the core side of the part becomes the inside (such as with a margarine container) of the plastic part.

Slide

Occasionally, a mold for a part cannot be designed in such a way that core and cavity pull apart because they will get caught on a component of the plastic part. The following illustration shows a snap fit feature, so called because the feature has to snap into another part. In this case, the feature would be torn apart were the mold designed as a separable core and cavity.

Plastic Tooling Descriptions and Terminology

Snap fit feature in a mold.

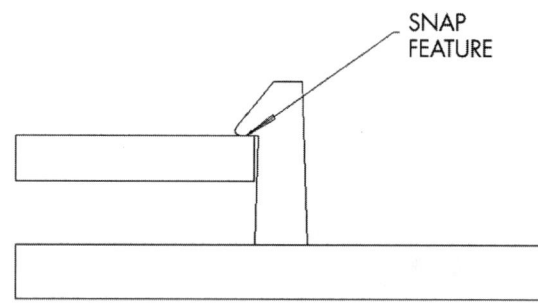

The snap fit feature shown in the previous illustration is such that the part cannot be removed from a mold designed for two solid pieces of core and cavity because the feature exhibits an undercut. The solution to this problem is to add a piece to the core that can be made to move within the core. In this case, the new piece in the core moves out of the way of the part before core and cavity are pulled apart. This new piece is called the slide.

Slides are parts of a core (sometimes of a cavity) that are used to create undercuts in the final plastic part. They are called slides because they "slide" out of the way of the plastic part when the plastic part is being ejected by the ejector pins. Otherwise, the plastic part would get caught in the mold and could not be released from the core or cavity. The following illustration shows a slide in the core of a mold.

Slide in core of a mold.

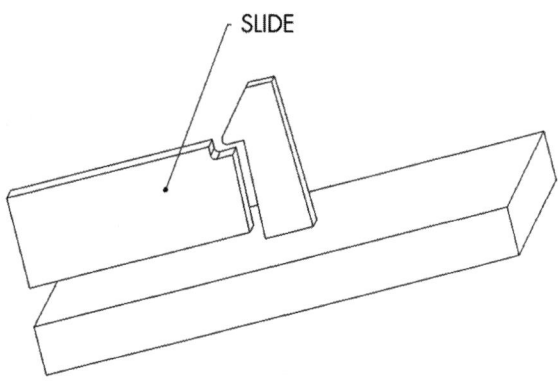

Features of a plastic part must be designed in such a manner that the slide does not interfere with any other features in the part. Adequate spacing of features must be made in the design for the slide to operate properly. The slide must also be able to move without getting in the way of the ejector pins or other parts of the mold.

A slide is usually found in the core of a mold because it leaves markings on the plastic part where the slide and core form the inside surface of the part. The slide must be slightly smaller than the space cut out for it in the core so that the slide can move within the core. This slight gap leaves an edge mark (or witness line) on the surface of the part. Because the user does not see the inside of the part, this is the ideal place to use slides so that they do not affect the cosmetic surface of the part.

Sprue, Gates, and Runners

When plastic is injected into a mold, there must be some way to transfer the melted plastic to the space between the core and cavity of the mold. A sprue is the opening through which the melted plastic enters a mold. If the plastic cannot be pushed completely into the mold through the sprue, additional paths need to be made for the melted plastic to flow to form the plastic part. These additional paths are called gates and runners. They distribute the melted plastic into the mold from the sprue. A sprue and runners and gates are shown in the following illustration.

Plastic Tooling Descriptions and Terminology

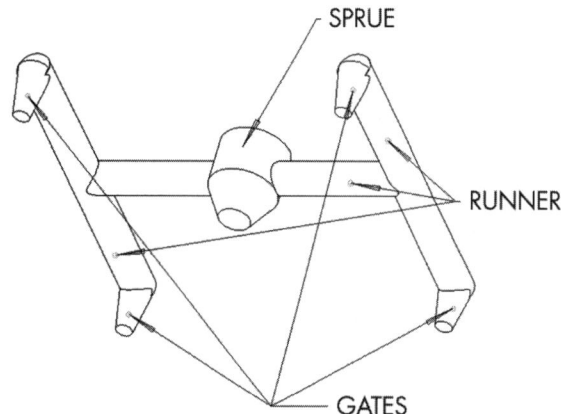

Sprue, runners, and gates.

The larger the runners, the easier it is for viscous plastic to flow from the injector to the mold. The more viscous the plastic, the larger the runners need to be to get the viscous plastic to the mold.

Runners and gates are used in other material molding processes. For molding metal products, the diameter of runners must be three to four times the size of those for plastic molding. In plastic molding, the material can be forced at a high pressure into the mold. For metal molds, the pressure is lower. Therefore, in order to get the viscous metal into the mold, a larger diameter is required for the metal to flow.

Cooling Line

In some molding processes, melted (viscous) plastic is forced into a mold. The melted plastic is hot and must be cooled to room temperature before the plastic part can be ejected. Time is needed to cool down the mold. A method to speed the cooling process is to have cold water flow into, through, and out of the mold. Cooling lines are designed into molds that handle this type of process. These lines, inserted or machined into the mold, direct the water to the parts of the mold that need to be cooled. The cooling lines must be placed to avoid the ejector pins, and close enough to the mold surface to adequately cool the plastic part. The following illustration shows cooling lines in a mold.

Cooling lines in a mold.

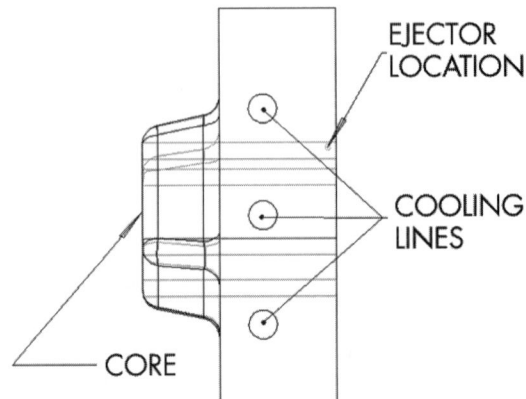

Electrical Discharge Machining

Normally core and cavity are cut from block with milling machines that use mills that have ball-type or square ends. However, sometimes these ends cannot create the square corners required for the final plastic part. In this case, a process called electrical discharge machining (EDM) is used, in which electrodes vaporize the metal to form the desired shapes of core and cavity.

Electrodes

In the EDM process, vertical electrodes are created that form the shape of the final design of the part. The electrodes might have square and sharp corners. Vertical and wire-cut are two basic types of EDM processes. The EDM process controls the sparks of electricity from the electrodes, which vaporize the metal in the core and cavity.

Wire electrodes are electrically charged, with the current traveling from one spool to another. As the wire passes near the metal core or cavity, the local metal is vaporized. This process is very similar to using a band saw to cut material. However, instead of a blade, electricity is used. Vertical and wire-cut EDM electrodes are shown in the following illustration.

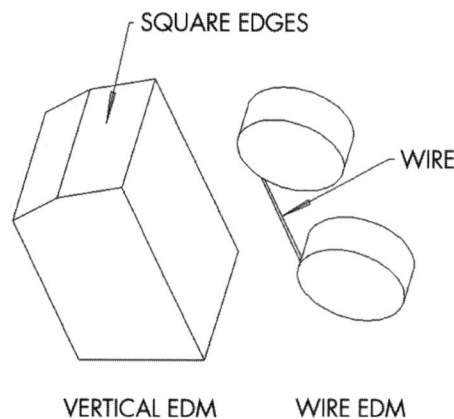

Vertical and wire-cut EDM electrodes.

Vents

Before material is introduced, a mold contains a volume of air. When the plastic material is put into the mold, the air has to go somewhere. If a mold were air tight, the plastic would go only as far as it could until the pressure of the plastic was equal to the back pressure from the air. This would leave the mold only partially filled (short shot). In this case, the compression of the air might also cause the plastic to burn and char at the locations where the air is compressed.

Vents are used to allow air to escape from a mold. Vent holes are typically small enough to allow the air to leave the mold but too small for the plastic to enter. Thus, molds are not air tight, but are "viscous material" tight.

This section has given you an understanding of the language of the plastic designer and mold maker. The next step is to see how these terms relate to the various fabrication processes.

Plastic Part Fabrication Methods

Many plastic part fabrication methods are available. The most common methods are described in the following sections. Each method requires a different plastic modeling technique. Therefore, the Pro/ENGINEER designer must adjust his or her design approach to match the manufacturing process intended for a given part.

Traditional and Thin Wall Injection Molding

Injection molding is one of the most common technologies used to manufacture plastic components in volume. It is used for fast part creation and low parts cost. The following illustration shows the significant parts of an injection molding machine.

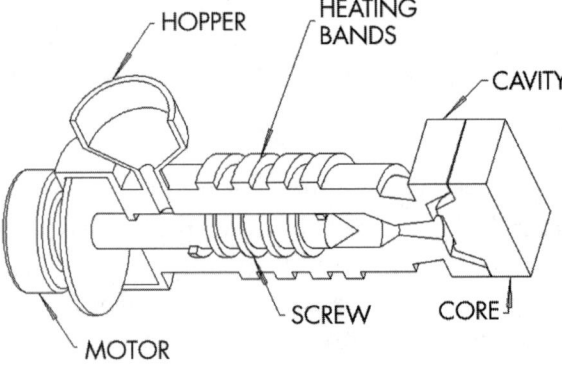

Injection molding machine.

The Process

Injection molding machines operate by having solid materials inserted at the start of the process and by having a plastic part come out as the result of the process. The solid materials are pellets usually made of virgin polymers. Sometimes crushed plastic parts of the same polymer can be reused as a percentage of the virgin polymers. The mixing of virgin polymers and ground, crushed plastic parts is usually done at the plastic manufacturing plant. This guarantees the user of the pellets a proper mixture of virgin and recycled plastics for the manufacturing of the parts. The pellets are about 3 to 5 mm in diameter.

Most materials must be kept "dry" before being put into a molding machine. Otherwise, the plastic will react with the water, causing the chemical bonds to change in the plastic as it goes from solid to liquid and back to solid. A poor or

Plastic Part Fabrication Methods

weak final product will result. The polymers are dried in an air-drying machine until the humidity of the polymer is at the supplier's specification. Today, this is a continuous system in which the pellets are dried and flow directly into the hopper. The purpose of the hopper is to hold the material and allow the material to flow smoothly into the heating section of the molding machine.

The heating section of a molding machine is called the injector. Electric heating bands in an injector heat up the pellets to the desired temperature. The temperature is high enough to allow the plastic to become viscous. The temperature cannot be too hot, however, or the plastic will burn, change color, or degrade into a physically weaker plastic. On the other hand, the temperature cannot be too cool or the polymer will not melt. This causes solid, non-melted polymers to go into the mold. The final part will have rough sections that did not bond.

The heating section also contains a space for the plastic about to be stored while it is hot, and before it is put into the mold. The storage size is set to hold the amount of plastic needed to fill the mold. This is called the shot size. If the shot size is too small, the mold will not be completely filled. This is called a short shot. If the amount of plastic stored is larger than the mold, the extra plastic is saved for the next shot. Other problems occur with an overly large shot, which are described in material to follow.

As the plastic, in its pellet form, flows from the feed hopper to the storage area, a screw device pushes the pellets along the barrel. The barrel is located between the screw and the heater walls. Here, the pellets heat up and become viscous. The screw is also a crucial part in the heating of the pellets. The screw produces heat by friction and shear stress. The shape of the screw allows the plastic to flow evenly toward the end of the screw. Screws are made in many different shapes and sizes, and can be changed so that the plastic approaches its ideal physical properties by the time it reaches the end of the screw. The pellets are totally viscous by the time they reach the storage area.

Within the storage area, the plastic is viscous and hot. The plastic then has to be pushed into the mold. The screw device is pushed forward so that the plastic in the storage area is pushed, or injected (injection molding), through the sprue into the mold, as shown in the illustration that follows.

Mold being filled.

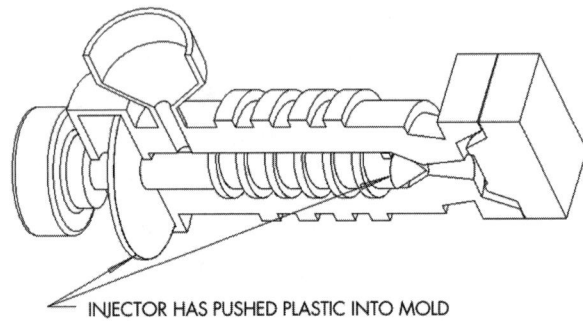

INJECTOR HAS PUSHED PLASTIC INTO MOLD

The plastic then fills the mold, with the air previously in the mold escaping through the vents. As previously stated, if there is not enough plastic in the storage area, a short shot results. The ideal is to have just a bit more material than is needed for the mold so that you have enough material to completely fill or pack the mold. However, too much plastic left in the overfilled area may degrade due to heating. Therefore, the molder must limit the amount of overshot.

As part size gets larger, larger and larger injection molding machines are required. The plastic is being injected into the mold under high pressure. The core and cavity of the mold need to be clamped together to resist the opening force of the plastic in the mold cavity. As the part gets bigger, the plastic in the core and cavity covers more surface area. Because the surface area is larger with bigger parts, more pressure is placed on the surface and more force is required to clamp the mold together.

A molding machine clamp is used to hold core and cavity together. The only force exerted on the clamps are the normal forces from the plane of the parting line or the projected surface area of the part onto the core and cavity.

For example, a test tube is long and cylindrical. The normal forces required to hold core and cavity together are those on the surface areas only at the end of the test tube. This area is very small compared to the overall surface area, as indicated in the following illustration. Most of the pressure from the plastic is on the core and cavity and not on pushing the core and cavity apart. However, if a simple plate were made, the normal forces would be much greater on the core and cavity because the area projected onto the clamps is much larger. The

Plastic Part Fabrication Methods

total surface area of the plate and test tube might be the same, but the normal projected area is much larger on the plate, with a higher clamping force required.

Force and ssurface areas related to test.

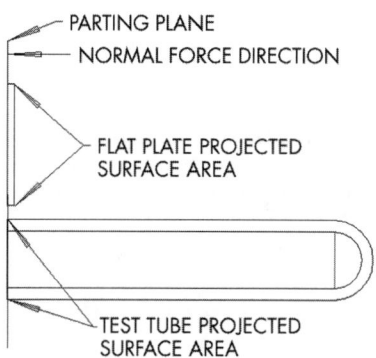

The force holding the mold together is typically measured as pounds per square inch (psi) pressure. The psi (or tonnage) is the rating given to injection molding machines. This rating tells the user that the machine can hold the mold together for a specified maximum projected surface area on the plastic part.

Most injection molding machines are computer controlled today. The pressure, temperature, and shot size are well controlled. Thermocouples are used to measure the temperature. Piezo-electric devices measure pressure. The stroke length of the screw device, at a given diameter, determines the volume of the shot. The operator sets all of these parameters so that the molding machine works at the proper and most efficient values. The mold opening of an injection molding machine is shown in the following illustration.

Mold opening.

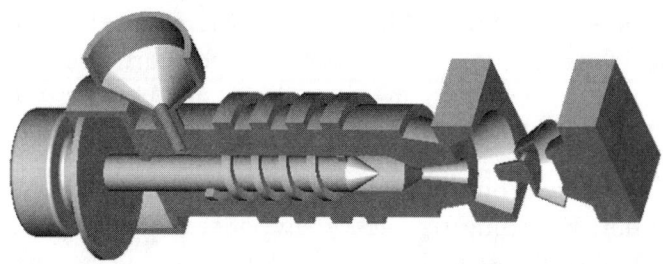

The cooling lines in the mold now cool the plastic. The temperature of the plastic is brought down to or near room temperature. The screw drive in the injector is backed off. As the mold is pulled apart, the ejector pins push the plastic part away from the core of the mold. The core is pulled away and the plastic part drops into a storage bin below the molding machine. The mold is closed. The process or cycle then starts over.

Cost Considerations

The injection molding process takes time. Injection molding machines are used for high-volume, low-cost products because they are expensive to purchase and operate. Therefore, the shorter are cycle times, the more plastic parts that can be made and therefore the lower is the cost of each plastic part.

An injection molding machine's cost of operation increases as the tonnage of the machine increases. This means that a 500-ton machine may cost $1,000 an hour to operate, whereas a 200-ton machine may cost $750 an hour. The designer has to be aware of these costs when designing a product. If the product can be designed to fit into a smaller molding machine, the per-part cost will be less, and the profit per part greater.

Another way of decreasing cost is to design multiple, similar cores and cavities in a single mold. This is called a multi-cavity mold. The plastic is injected into all of the cavities at the same time using sprues, gates, and runners. The user gets many parts in one cycle, thus reducing the cost of each part compared to getting one part per cycle.

For example, if a one-cavity mold is used in a 200-ton machine at $750 an hour and the cycle time for a part is 30 seconds, the cost for one hour's use of the machine is $6.25 per part. If a two-cavity mold is used in a 500-ton machine at $1,000 an hour and a cycle time of 30 seconds, the cost for one hour's use of the machine is $4.17 per part. The conclusion is that it may be more economical to use a multi-cavity mold on a higher tonnage injection molding machine.

Another method of decreasing per-part cost is to make the walls of the part thinner. By doing this, less material is required for the mold and the part, and the per-part cost is decreased. However, the design of the part's features must

Plastic Part Fabrication Methods

be modified. (This is discussed in later chapters.) The injection molding machine must also be changed. The walls are thinner; therefore, it will take a higher pressure to push the plastic into the mold. More gates and runners will be needed to push the plastic into the mold.

The cost of the molding machine goes up because more features are needed in the mold. A higher tonnage injection molding machine will be needed to withstand the packing pressure of the plastic. At the same time, less material is in the mold, reducing cooling time. This reduces the cycle time. A lower cycle time means more parts per hour, thus reducing the individual cost per part. All of these factors must be taken into consideration to determine if it is economical to go to thinner walled products.

An important fact of which the designer needs to be aware is that different plastics shrink at different amounts. The shrink percentage may vary in the x, y, and z directions in the part. This means that the same mold cannot be used for different plastics because the final plastic part will not be the same size. The designer must make the molder aware of the material to be used in the final part so that the mold can be made larger to account for the shrinkage of the material.

Gas Assisted Injection Molding

Gas assisted injection molding, shown in the following illustration, is an enhanced process. The main aim of this process is to reduce the amount of material used to create the plastic part. In standard injection molding processes, the plastic fills the entire mold. This process works well if the wall thickness is constant. However, if the part has thick and thin sections, the thick sections shrink more than the thin sections. This extra shrinkage appears as sink marks or indentations in the part.

Gas assisted injection molding.

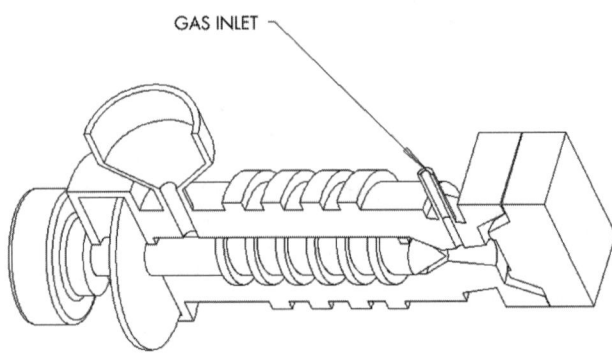

The previous illustration shows a gas injection molding machine. The plastic is injected into the mold but does not completely fill the mold. High-pressure gas is immediately put into the mold. This gas flows into the lower pressure sections of the mold, which happen to be the thicker sections of the design, thus hollowing out the thick sections of the part and pushing the remaining plastic into the unfilled sections of the mold. An example of a part made by gas assist injection molding is shown in the following illustration.

Cross section of a part made by gas assisted injection molding.

The part shows where the gas has made the walls thinner in the thick sections. By doing this, no sink marks are formed by excess shrinking of thick sections. The internal stress in the part is lower because there is reduced injection pressure to pack the mold. Therefore, the part has a smaller chance of warping once it is removed from the mold. This gives the designer a more durable and repeatable product. Here, less material is used, and the cost and weight of the product are reduced. Also, the cycle time is reduced because the part can be cooled faster due to thinner walls.

Plastic Part Fabrication Methods

The disadvantage to this system is that the mold must be made of specially tempered steel. This makes the tooling cost higher than for other materials. Mold flow analysis must be performed to ensure proper filling of the mold. The design of parts must allow for the flow of the plastic and gas in the thick sections of the mold. The thin sections must not cool and solidify before the gas can get through to the thicker sections of the part.

Foamed Parts

Foamed molding is used for large parts that require structural strength but are not necessarily high-volume products, such as garbage containers. These parts usually have relatively thick walls, for strength, which allow open or closed bubbles to exist in the foam. The following illustration shows an example of the molding machine used for structural foam injection.

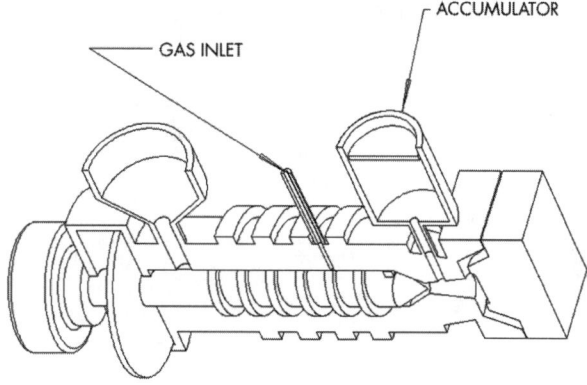

Structural foam injection.

A structural foam injection process is very similar to the standard injection molding process. There are two methods of creating plastic parts with the structural foam injection process. In the first method, plastic pellets are placed in a hopper and fed smoothly into the screw in the injector. The screw turns and the pellets are moved down the screw. The pellets are melted into a viscous state. At this time a gas is injected into the screw. The plastic and gases are mixed in an accumulator in the shot area. When the proper shot size has been stored in the accumulator, the accumulator then pushes the mixture into the mold.

The second method uses chemical blowing agents, which are added to the plastic pellets. The pellets are placed in the hopper and fed through the screw of the injection molding machine. The blowing agents absorb enough heat to expand, but cannot expand due to the pressure in the screw and accumulator. The plastic is then injected into the mold and the gas expands inside the mold and starts its foaming process.

As the plastic comes in contact with the mold, it cools down, forming a thin, solid surface. Cellular (closed) bubbles are formed between the cavity and core surfaces because the pressure is lower in cavity and core than in the accumulator or screw, thus allowing the gas to expand. This produces a plastic part with solid surfaces and a cellular center. An example of a structural foamed part, a handle, is shown in the following illustration.

The handle has two purposes. It is lighter than a solid plastic part because it contains less material. It is also stronger than a hollow plastic part because the cellular structure is stronger than a single cavity.

Cross section showing structural foam used to create a handle.

Extrusion Molding

Extrusion molding is used for shapes of uniform cross section. Examples are eaves troughs and window seals. The following illustration shows the cross section of a part that has been extruded.

Plastic Part Fabrication Methods

Cross section of an extruded part.

Extrusion molding is different from standard injection molding for two reasons. Unlike injection molding, the process is continuous, and a die is used instead of a mold. An extrusion molding machine is shown in the following illustration.

Extrusion molding machine.

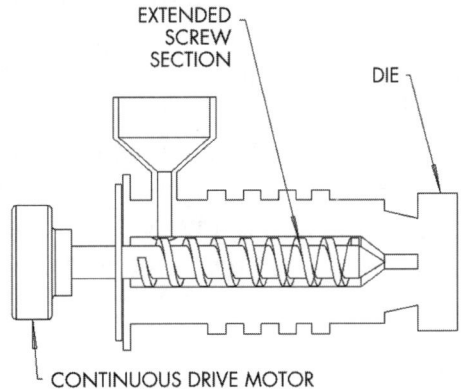

In the extrusion molding process, plastic pellets are placed in a hopper. They are fed smoothly into the injector section of the molding machine. The screw of the mold machine is always turning and always pushing the plastic through the molding machine. The pellets travel along the screw and are heated by the electric heater bands around the screw. By the time the plastic gets to the end of the screw, the plastic is in a viscous state. The plastic is then forced into the die.

The die represents the mold of this type of molding machine except that it does not consist of a core and cavity. Rather, it is the shape of the final part. The die contains cooling lines, which cool the plastic as it flows through the die. The plastic is in a semi-solid state when it leaves the die.

> **NOTE:** *Chapter 12 presents an example of using this semi-solid shape to advantage in making parts using Pro/ENGINEER as the design tool.*

Blow Molding

Blow molding is used for parts that cannot be designed with a core, but do need a void in the center of the part. Some examples are bathroom shampoo bottles and toy balls. If these parts were made in a standard injection molding machine, the core would be impossible to remove. Instead of using a core and a cavity, blow molding uses two cavities that form the final product when put together, with a parting line typically visible where they join.

This book covers the basic ideas involved in the process of blow molding. However, the production of soda pop bottles, for example, uses both the injection molding and blow molding processes. The combination of the two processes will not be described here, but the reader should recognize that some plastics could be reheated and reused in different processes to arrive at a final product. The designer must be aware of all processes and use all of them to create the best product for the market. Blow molding is done in three stages. The first stage is shown in the following illustration.

Plastic Part Fabrication Methods

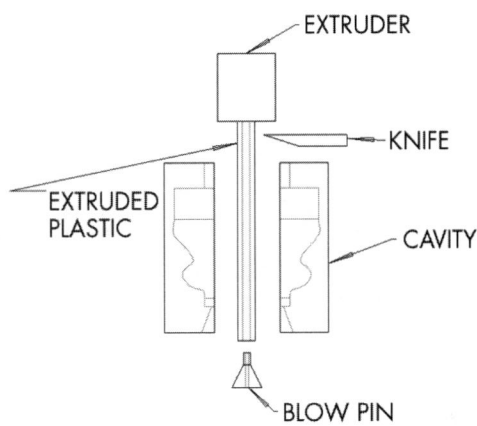

First stage of blow molding.

The molding process is done vertically and makes use of the force of gravity. The injector in this case is like the extrusion molding machine except that the flow of material is not constant. The die used in this process is normally circular. The plastic that flows through this die flows out in the shape of a tube (or parison).

In blow molding, the plastic material is not completely solid when it leaves the die. That is, the plastic is still flexible. For this reason, the tube of plastic must be vertical and use gravity to keep its tubular shape. If it were horizontal, the tube would bend and the process would not work.

A shot of the plastic is pushed through the die, assisted by gravity. The tube formed as a result sinks due to gravity to a blow pin. The purpose of the blow pin is to put a gas into the tube and have the tube expand. This second stage of the blow molding process is shown in the following illustration.

In the next stage, as shown in the following illustration, the mold cavity closes. As the mold cavity closes from both sides, the top part of the plastic tube is cut and sealed off by a cutter mechanism. The purpose of this is to hold the plastic tube in the center of the mold and to prevent any gases to exit through this section of the tube.

Second stage of blow molding.

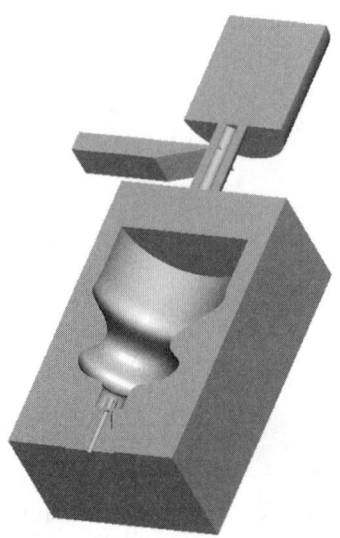

Once the cavity is closed from both sides and the tube is cut and sealed at the top, gas is applied through the blow pin. Because the gas cannot escape through the top side of the tube, the gas forces the tube to expand. Remember that the plastic is still flexible and can still be formed. The tube expands due to the gas pressure until it touches the wall of the cavity. The wall of the cavity is cool, which causes the plastic tube to solidify and take the shape of the cavity. The tube expands until all of its sides have come into contact with the cavity and have solidified.

The third and final stage of blow molding is shown in the following illustration. The blow pin is moved out of the way, and the two cavity walls are separated. The part is then removed. The top section of the part, where the original plastic tube was located, is also removed.

Plastic Part Fabrication Methods

Third stage of blow molding.

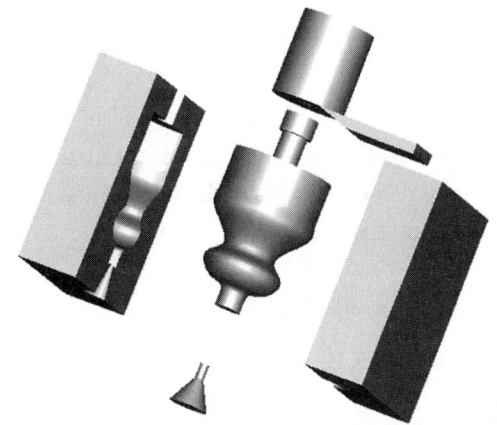

If you look closely at plastic bottles, you can see the parting line, formed by the two cavities, on the sides of the bottle. At the bottom of the bottle, you will see where the extra material from the original tube was cut off. In addition, the wall thickness is different due to the gravity and the distance the tube of plastic had to be moved from its original position. Designers have to be aware of this when designing parts for blow molding. Walls can be too thin for use with this process. The plastics used in this process can be opaque or clear. It depends on whether the final customer wants to see what is inside the container. The plastic materials used can be flexible or stiff once the process is complete.

This molding process is relatively inexpensive. Fine detail, which would add significantly to cost, cannot be designed into the cavities because the plastic from the original tube would not fill in these areas. Not enough pressure is available from the blow pin to push the plastic into detailed areas. In addition, too high a pressure may cause the tube to burst before it expands to the cavity walls. Therefore, the cavities are easy to manufacture because of the simplicity of the design and machining requirements.

Rapid Prototyping and Tooling Methods

The processes previously mentioned require 6 to 16 weeks to produce the core and cavity for the molding process. These molds are used for production manufacturing in which thousands and maybe millions of parts can be molded. Because of the time it takes to produce molds, they are often very expensive. In addition, if changes to a mold are required because the parts are not quite right, the changes are very expensive at this stage in manufacturing.

Rapid Prototyping

To avoid the expensive cost of changes in the final mold, rapid prototyping processes for creating parts have been developed. The purpose of these processes is to take the CAD file, for example a Pro/ENGINEER file, and convert it into one real part. The part is rarely made of the same material as the material required in manufacturing. This means that the part will give the designer the form and fit of the final product, but will not give the designer a truly functional part.

Prototyping parts can be made of plastics (i.e., stereolithography), paper (i.e., laminated object modeling), metal (i.e., metal deposition), or even edible starches. The parts are useful because they can often be made overnight, and design faults can be fixed. Any design changes can be made before going into full production of the manufacturing molds.

Ninety-five percent of the rapid prototyping processes involve a 2D layer creation method. The processes create the 3D part by putting each subsequent layer on top of the last. Some methods use lasers to create each layer by adding material to or removing it from the previous layer. Some methods use x-y plotting methods to add material to a previous layer.

The only commercially available method at the time of this writing that can create 3D shapes of materials that can be used as molds for injection molding machines is computer numerically controlled (CNC) milling. However, this method is probably the slowest of the rapid prototyping processes. The mills

Rapid Prototyping and Tooling Methods

used in CNC cannot make the sharp edges often required in a final part. In addition, if the part has an undercut, the part usually cannot be made as one solid piece using CNC.

Rapid Tooling

Rapid tooling is a method of creating a mold faster than can be done by conventional means. Rapid tooling, which often uses rapid prototyping parts to create a tool, can significantly reduce the time it takes to make a tool.

Most often the number of parts that can be made from these tools, which are created as molds, is less than 50. If the designer needs prototype parts that are very close to the physical properties of the final product, the designer may choose to use this tooling process to obtain those prototype parts.

Epoxy Molds

Epoxy molds are used to quickly make the core and cavity of a mold. The mold will make 10 to 20 parts. In this process, a part, which is usually made by a rapid prototyping process, is held in a container, as shown in the following illustration. The part is held away from the walls of the container.

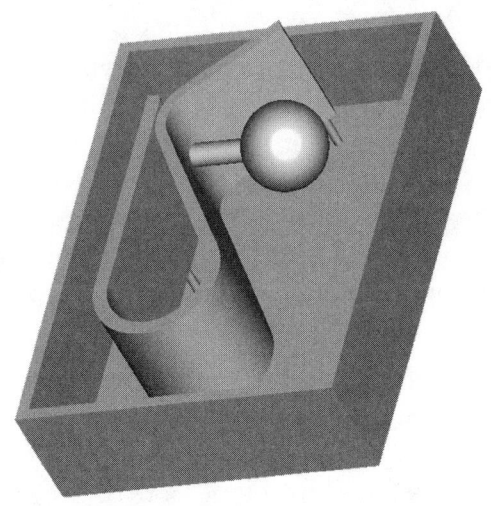

Part in container for rapid molding.

An uncured epoxy material is poured around the part until the level of the epoxy material is up to the parting line of the mold. The epoxy material is allowed to cure and solidify. A material (called a mold release, which prevents adhesion of similar materials) is sprayed over the mold and prototype part. This is done so that the next layer of epoxy material may be removed from the first layer of epoxy material. The next layer of epoxy material is poured into the container and allowed to solidify.

The first and second layer of epoxy are then taken apart at the parting line. The prototype part is removed. The two epoxy parts left are the core and cavity of a mold of the prototype part. The core and cavity can now be used as a mold for a more realistic plastic part.

In this process, the plastics (urethane casting, for example) used to create the final parts must be in a fluid state to be able to pour into the mold. The plastics must then cure in the mold. Heat from exothermic reactions as the plastic changes from a liquid to a solid must not be too high or the mold will be destroyed. To keep the heat at a workable level, the reaction of the plastic must take place over a long period—about 2 to 6 hours. The heat from the plastic changing from liquid to solid does, however, eventually degrade the epoxy mold. This is why only 10 to 20 parts can be made from the mold.

Plastic Molds

Recent developments in rapid prototyping have made it possible to use prototype parts as molds in injection molding machines. Instead of fabricating the prototype part from the part in the rapid prototyping machines, the core and cavity are fabricated. Using Pro/ENGINEER, for example, the part CAD file is used to make a core and a cavity CAD file. The core and cavity files are modified to allow for shrink, cooling lines, sprues, runners, and ejector pins. The rapid prototyping machines then fabricate the core and cavity from the CAD files.

The prototype core and cavity are placed into an injection molding machine. The molding machine is run, but at a very slow cycle. Each cycle may be 5 minutes long as opposed to 15 seconds. The purpose for the long cycle time is to give time for the prototype core and cavity to cool down. Otherwise, the core and cavity get too hot and melt. This process has been proven to produce up to 50 parts in the desired plastic material.

Summary

Some of the basic terms for molding machines and processes have been described in this chapter. You should now have a basic understanding of the terms and principles related to the area of plastic molding. Many molding options are open to the designer, whose task it is to decide which manufacturing method best suits the final use of products and their required quantities. The next step is to understand why you might select plastic instead of metal as the material to be used in manufacture. The next chapter explores this selection process.

Chapter 2

Why Select Plastics?

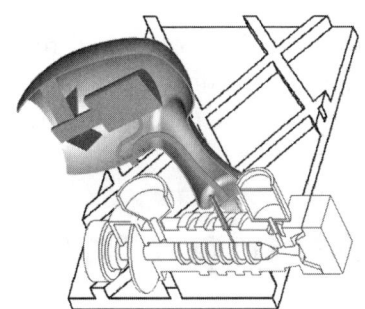

Introduction

Plastic technology is commonly used in design because of the versatility of the material properties of plastics. Using plastic as your material of choice allows you to design for a cosmetically pleasing yet functional component of the highest quality. Plastics also lend themselves to mass production of final products with predictable results.

Plastics are increasingly used to replace parts traditionally made out of metal. Metal materials are usually heavier than plastic materials. The heavier a product, the greater the cost of its transportation to the marketplace, and the greater the expense of energy consumption involved in its use. For example, the weight of cars has been greatly reduced by using plastics instead of metals. This reduces the weight of a car, making it more fuel efficient and less expensive to market and operate.

Plastics also offer the advantage of ease of creation where a metal part of the same configuration would be difficult to achieve, and probably more expensive. Plastic parts can be made into many shapes easily. Metal shapes can often be difficult if not impossible to make without expensive tooling. Tooling for molds for plastic parts can also be expensive, but the trade-off in the choice of material may be in the additional functionality and cost savings plastic might give over metal.

Plastics can also be used to replace many metal parts with one plastic part. An example would be the button on your calculator. It is one piece of plastic and it replaces many metal parts such as springs, caps, and metal electrical contact pieces.

Electrical Insulation Properties

The energy output of electricity varies greatly in terms of its use. You flip a plastic protrusion on a light switch and the light in the room goes on. During thunderstorms there is lightning, which is electricity with a huge amount of short-term current and an extremely high voltage. Electricity powers your vacuum cleaner, as well as your hand-held computer, and even the greeting cards that play tunes to you when you open them. All of these forms of electricity and its use have heat (energy conduction) factors associated with them.

Pure plastics, at room temperature, are electrically nonconductive. Metals, at room temperature, are conductive. Ceramics, at room temperature, are nonconductive. The reason for mentioning temperature is because it the most important factor in choosing a material for electrical shock prevention. At extremely low temperatures, for instance in outer space use, ceramics may become conductive, and metals may become superconductive. In outer space, temperatures can be extremely low. Therefore, for space applications, plastics may be the only solution for insulating electrical components.

Plastics are often the material of choice to protect the user of a product from any shock or hazardous effects that may come from electricity. Often plastics are used to hide electrical conductors and motors, such as in a hair dryer (shown in the following illustration), and to protect you from the electricity at the power source, such as your household outlet.

Electrical Insulation Properties   37

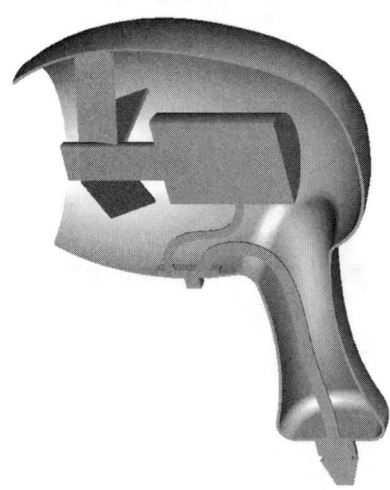

Cross section of a hair dryer.

You can see from the previous illustration that the plastics form an insulating layer between your hand and the electricity that drives the motor. The main factor for insulation against electrical shock is to provide a separation path between you and the electrical device. The driving force in electricity is the voltage. The lower the voltage, the shorter the separation path needs to be. For instance, in the wintertime, when the air is dry and cold (not good for conduction), you might note that you can easily get a friction "shock" (high voltage) from, say, touching a piece of grounded metal. The minimal separation path to avoid the shock in this case might be just two centimeters.

In the case of the "friction" shock, the electricity has to build up enough energy to break down the insulation of the air and transfer this energy from you to the piece of metal. You can compare this to a dam two meters high holding back one millimeter of water. Once the water reaches the two-meter height, the dam breaks and the water falls, releasing a great amount of potential energy (voltage) but for a very short time (milliseconds). This compares to a dam holding back a two-meter height of water that is a kilometer in length. When the dam breaks, the water falls, creating a great amount of potential energy (voltage). In this case, however, the potential energy is created for a very long time because it takes time for the backlog of water to flow out. However, with a spark or shock, there is no reserve of potential energy. Therefore, you have a tall column of potential energy that is changed to kinetic energy very quickly, which then has no potential energy to continue the flow of energy to a source.

In the case of the "friction" shock, if you add an insulation barrier, such as a plastic material, that increases the length of the path to more than the two centimeters the electricity has to travel, you would not get an electrical discharge and therefore no shock. The path is too long for the potential energy to cross the insulating effect of air and plastic. The following illustration shows an example of the path electricity would have to take around a piece of plastic in order to cause a spark.

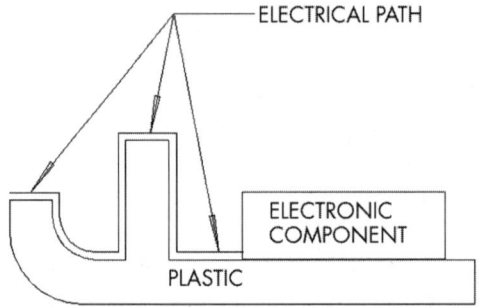

Electrical path around a piece of plastic.

When designing plastic parts for electrically powered equipment, the designer must be aware of the paths any electrical discharge might take. Two good examples are the space along the parting line of the parts and the space between a display module and the cover of a part. The designer must be sure that electricity cannot find any path to the user of the device. You do not want to electrocute the user and you do not want the user to blow out the electronics by sending a static discharge to them.

A rule of thumb to use for preventing static discharges is to use 1 millimeter of separation for every thousand volts of discharge. Therefore, if you are making a product that will be used in the winter in a dry area, such as the north central plains of the United States, the path length should be at least 20 millimeters to stop a 20,000-volt spark from damaging your product.

Although plastic is an insulator, it will only insulate if it is thick enough. A piece of plastic that is paper thin will not prevent a spark from going through it. This is because plastics are not completely solid in their chemical bonds (that is, of the molecular chains that make up the plastic). Minute air lengths will be found in all plastics. With thin-walled plastics, the electricity can find a path length that will allow it to go through the plastic and not around it. Therefore, the designer must be aware of possible static discharges through any thin-walled plastic area. A good way of determining the minimum thickness of a material is to use the dielectric constant of the material. This value determines the minimum thickness a material needs to be to inhibit a spark.

Cosmetic Appearance

The use of plastics in design allows designers to select from a variety of cosmetic appearances while maintaining the intended function of the component. The visual appearance of a product obviously affects its marketability. A product that is square in shape might not sell as well as a similar product that has flowing lines and smooth contours if the product is intended, for example, to be held in the hand. The square mouse for a computer, shown in the following illustration, does not look the least bit appealing when compared to a mouse that has sculptured surfaces. They both do the same thing electronically, and they both cost the same, but the user will probably buy the mouse that is smoother, more attractive, and more comfortable to the touch.

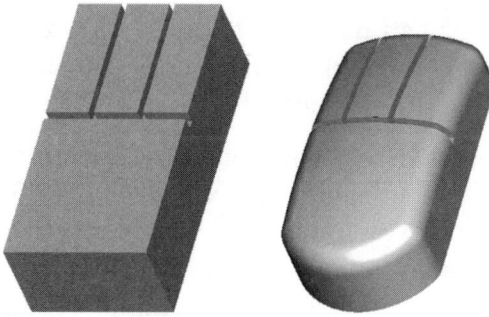

Comparison of a sharply shaped product with a sculptured-surface product.

Surface texture also affects a product's appearance. The designer would like to have all the surfaces smooth. This gives the designer a chance to keep the draft angle down to a minimum and make the job of designing easier. However, smooth surfaces tend to highlight any defects in the design of the part and any faults in the molding process, such as sink marks. Smooth surfaces also tend to show fingerprint smudges and scratches.

> **NOTE:** *Draft using Pro/ENGINEER is discussed in Chapter 5.*

In other cases it might be desirable to include smooth surfaces in a design. For example, a designer might want a particular portion of a design to stand out, such as for a label, or a button that needs to be visually obvious against a textured surface.

A textured surface can make a product commercially acceptable but often makes the design of plastic parts more difficult. A textured surface is a surface that has tiny bumps on it, as compared to a smooth surface that has a specular reflection like that of a mirror. You could compare the types of textures of surfaces to the textures of sandpaper. The texture might be very rough and grainy, like number 40 sandpaper, or very fine, like number 1,000 sandpaper.

Textured surfaces are also often not uniform. That is, they may contain decorations, such as a logo or the pattern of a brick wall. Plastics designers do not necessarily create these decorations, but indicate to the mold maker how to incorporate such patterns in a mold, such as where to position a pattern for the intended aesthetic effect. This patterning process is called etching, which the mold maker performs as part of the mold creation process.

The plastic part that has a textured surface often hides sink marks. The textured surface also hides plastic flow lines that show the direction in which the plastic went when it was injected into the mold.

The plastics designer has one problem in designing when textured surfaces are used. The draft angle may vary from 2 to 5 degrees, depending on the coarseness of the texture, giving the designer less space with which to work inside the design. The overall part dimension may need to be changed because of the loss of internal space. The designer might well have choices to make or advise on in

Versatile Part Design of Complex Parts

regard to changes necessary to the curvature and cosmetic appearance of a part design to keep the final product aesthetically pleasing and functional.

> **NOTE:** See Part III for "hands on" Pro/ENGINEER exercises.

Versatile Part Design of Complex Parts

Plastics are used to create very complex components. As previously mentioned, it is common to have one plastic component replace several other components in a product design. As was stated in the introduction, a plastic button can be used to replace a button that has springs and other materials. The following illustration shows an example of a plastic button that has replaced a multi-part button.

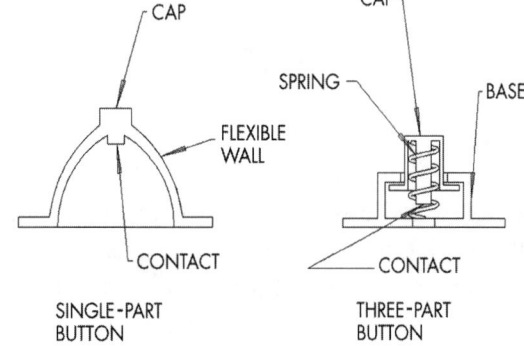

Cross section of two button designs.

The single plastic part in the previous illustration performs the action of a spring. The multi-part button has many parts that need to be made and assembled, which cost money. Therefore, the single button is less expensive and uses less material for the design objective.

An example of a very complex design would be the frame mechanism on the inside of a car door. This mechanism would hold speakers and the sliding structure for the window. The frame could be made of many metal parts that have to be stamped and then welded. However, there are several considerations that might make the designer opt for a material other than metal: (1) the metal parts would be heavy, (2) the tolerance allowances of fitting all of the parts together would be a designer's nightmare, (3) the frame would have to conform to the outside shape of the door, and (4) the frame would have to avoid any electrical parts providing power to the windows and speakers.

In this case, a plastic could be used to replace the metal structure of the frame. The advantages are numerous for this particular part. Plastic would be as strong as the metal frame. It would be lighter, reduce weight, and eliminate unnecessary handling of individual components. A simpler design would maintain functionality. Because the frame would consist of one piece, the need for tolerance analysis would be eliminated. The engineering analysis of the product would also be reduced because only one part is being analyzed for vibration and strength. The plastics designer needs to be aware of replacing parts consisting of numerous components with minimal plastic parts that do the same job.

Flexible Material

Plastic material is also used to great advantage in applications for which flexibility is desirable, such as in the case of hinges and springs. With such parts, the designer needs to design and incorporate parts that can flex without breaking or snapping. An example of such a part is a button for a switch or a hinge for a cover, as shown in the following illustration. In the case of the hinged cover, the designer wants the door to stay in place and not flop around, yet at the same time the door must be able to bend 180 degrees.

Flexible Material

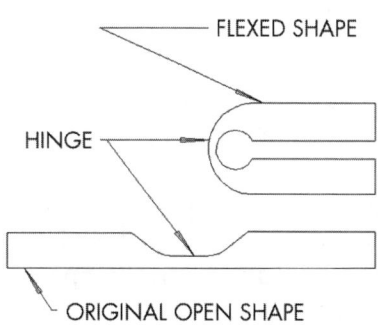

Cross section of a door with a flexible hinge.

The "live" (flexible) hinge works because the material is very thin at the location the designer wants the material to bend and flex. Plastics are often preferable to metals for such applications for the following reasons. Metals have linear stress/strain curves, which determine at what point and with what force you can bend a metal. Most metals can be deflected to a point beyond which they will not return to their original position. For requirements of high deflections, only certain metals can be used.

The problem with these metals is that they are only good for one function; for example, as springs within a clock. The outside of the clock would be made of a different material. Plastics in this case show their versatility. Plastics have nonlinear stress/strain curves, and some plastics can accommodate large displacements and return to their original position. For "live" hinges to work, the plastics must be able to withstand extremely large deformations and still be able to return to their original position. At the same time, the same plastic can be used as the cover of a part. This gives the plastic two functions in the same part, which is often not possible with metals.

Plastics Properties

Only special types of plastics can be used for the purpose of withstanding high deformations. The plastics must have a molecular structure that has long chains of molecules. Each of the long chains of molecules is loosely bonded. The chains can then move among themselves without tearing the bonds that hold the chains together.

Plastics also have a "memory." That is, once they are deformed, plastics tend to want to return to their original position. However, it should be noted that in any deformation a certain level of dislocation of molecular structure occurs that cannot completely be recovered from. That is, the plastic cannot completely return to its original shape, even though the dislocation is unnoticeable.

Depending on the plastic, the designer can design a part to flex for ten times or up to millions of times. The start of the failure of a live hinge, for example, can be seen when the hinge starts to appear white and the material gets thinner at a particular area.

The Pro/ENGINEER designer can use the flexing property of plastics to make special features (see previous illustration). Sometimes metal is the material of choice for the bulk of a part, but the metal does not have all material properties needed to perform the functions required in the product. In these situations, plastic parts can be combined with metal parts to supply the functionality.

Plastics can be used to reduce the number of parts necessary in a product and at the same time supply functionality that may not be possible with metal. An example is a snap fit, as shown in the following illustration. The plastic hook passes by the lock feature when the parts are assembled. Once it has passed the lock feature, the hook snaps back to the original position. This type of feature is used to hold two parts together, and is often used in situations in which the functionality it provides is needed once in the life of the product, or so infrequently that wear and breakage are not issues.

Snap fit.

Snap Feature

If the plastic hook in the previous illustration were metal, it would have to be mounted onto the plastic base. The plastic hook is performing the same function a metal hook might, and the plastic hook is probably near its deflection limit. However, the plastic hook and base are one part, whereas a metal hook could not be one part with the base. There are perhaps spring-type metals that could perform the same function, but the designer would be adding parts to a product, or making an assembly. This increases the cost and number of parts in a product when compared to one plastic part that does the same thing.

Other Material Property and Design Considerations

Other material property considerations include energy absorption, deflection resistance (strength), surface durability, part shape (such as functioning button design), and associated costs. These issues are explored in the sections that follow.

Energy Absorption

Designing products in plastic requires that you give special consideration to the selection of materials, as well as to the overall function of the component. Pro/ENGINEER allows the designer to quickly add features to a design, which aids in producing a robust yet functional overall design. The following example of adding a round corner to a square corner can be done in seconds in Pro/ENGINEER.

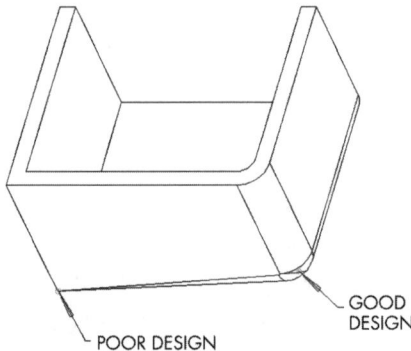

Box with a good and bad design for corner shapes.

Almost all products are made to withstand some type and level of bumping, dropping, or moving around. Every time a product is moved, it picks up kinetic energy. When the product stops moving, the kinetic energy has to go somewhere. The kinetic energy is transformed into other types of energy, such as sound, heat, entropy, and potential energy. The product must be able to absorb or transmit this energy or it will break.

An example of this transference of energy is a ball that drops on the floor. It makes a noise each time it hits the floor, giving off sound energy. It also compresses and stores the energy as potential energy. As the ball compresses, its temperature rises, thus absorbing energy. It releases the potential energy by going back to its original shape and rising in the air. It does not go as high each time it falls to the floor and bounces back up because of the energy lost in the heat rise and the sound energy expended. Air drag also reduces the energy available for the ball to absorb each time. Eventually, all of the energy is dissipated into heat, noise, and air drag and the ball comes to rest.

Products must often be designed to withstand being dropped or moved without breaking. That is, they need to a certain extent to behave in the same manner as the ball. Features can be added to a part to help prevent breakage. For example, the previous illustration shows a sample design of a box with one square corner and one round corner. If the box is dropped on the square corner, most of the energy will be absorbed in the corner. This is where a stress concentration will occur. Plastics and metals break or fracture at stress concentrations. Therefore, it is likely the part designed with a square corner will break at the corner if dropped.

Other Material Property and Design Considerations

The designer can fix this problem by adding a curve to the edge of the part, as shown on the other side of the box (see previous illustration). The designer is removing stress concentrations. The part will transmit the potential energy throughout more of the part and hopefully no areas of the part will have any stress concentrations. The designer can further reduce any chance of stress concentrations by adding rounds to the part at any sharp interfaces of edges.

> **NOTE:** *Examples of adding rounds are found in Chapter 5.*

The designer may also add interlocking features. These features allow the cover and base of a part, for example, to slide along each other without the part completely separating at the parting line. The following illustration shows an example of this type of feature.

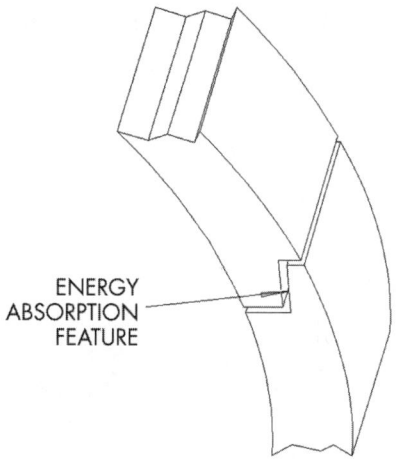

Feature for aiding in energy absorption.

The designer may also have the opportunity to select form various types of plastics. The plastic chosen might be flexible, absorbing energy and bending as the energy is transferred down the part. Conversely, the designer might select a fiber-filled material or a material similar to polycarbonate, both of which would produce a stiff part exhibiting low properties of deflection for use in a situation where this is desirable.

In the case of a stiff material, the energy is transferred to the features holding the part together. The overall shape of the product does not bend or warp, which in some cases might damage parts inside the product. The point is that the designer must determine the necessary characteristics of the material used for any given part or combination of parts based on intended use, durability, and other factors.

Deflection Resistance

Pro/ENGINEER provides designers with quick methods of creating features that add to the overall stiffness of components. Flat, thin pieces of plastic, for example, are very flexible. Flat pieces are often found on covers of electronics equipment such as those for printers, stereos, and personal computers. Such parts often require features that will add structural strength. In this case, such features are called ribs.

Pro/ENGINEER is one modeling package that allows designers to easily make rib features. The rib features add strength (structural rigidity) to the part. The following illustration shows a part with rib features.

Part with rib features.

Ribs do not have to be parallel to the edge of the part. They can be at any angle. They can also cross each other to help retard the twisting and warping of the part. Another option for designing parts that withstand deformation is to make the surfaces of a part "sculptured." This is done by adding tweak pushes and pulls on the surface. This creates concave and convex areas on surfaces. When a force acts on a surface, the concave and convex areas tend to mitigate against the adverse effects of deformation. The forces on the surfaces will be compression and tension forces on the surface and the part will tend to stay in its original shape.

Durable Surface Finish

Some products experience wear due to friction caused by the movement of another object against them. The object could be a finger, a piece of plastic that is part of the product, or an instrument used to perform some operation on the plastic, such as pushing a button with a pencil. The designer must know of materials and modeling methods that will help prevent the wear of a product.

Plastic materials can be chosen that are scuff resistant. These materials have hard surfaces and are very rigid. The designer must locate these parts where flexibility and stress concentrations are not significant to the product. Otherwise, the product will break or fracture under shock due to the inflexibility of these areas.

An option for the designer is to make the part smooth, because friction is reduced on smooth surfaces as opposed to textured surfaces. Textured surfaces tend to wear faster and more noticeably with time, whereas smooth surfaces retain a look that is closer to the appearance of the part when it was new. The surface might be concave or convex, as indicated by the buttons shown in the following illustration.

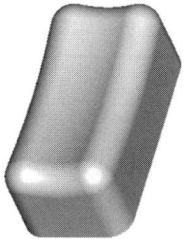 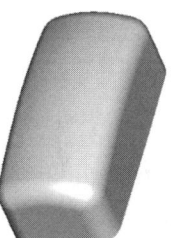

Smooth-surfaced concave and convex buttons.

Functioning Part Design

On a flat surface, an object pressing on the surface will have a tendency to slide. This causes wear on the surface. However, the finger that pushes the button on a concave surface will tend to rest in the button and not slide around, thus reducing friction and wear. In a convex button, the finger tends to rest on the first part of the button it touches and reduces the chance of the finger moving around. These are two design considerations in reducing product wear. Another option for preventing wear, in this case for preventing a finger from creating friction, is for the surface to be textured. This can be done with labels (such as the symbols on the buttons of a telephone) that are indented instead of protruding in a product surface. The following illustration shows an example of button labeling.

Button labeling.

By making the letters indented (cut into as opposed to stamped, printed, or raised), the lettering will not wear away. The opposite occurs for features made with protrusions. These features have a tendency to wear, sometimes becoming unrecognizable in a relatively short period of use.

Associated Costs

Plastic is typically selected in designs for its adaptability to a low-cost approach to the overall design. Plastic products often have fewer parts than products made with metal. By having fewer parts, the total cost of manufacturing the product is reduced. Other factors that enter into the design decision are:

- The assembly time is shorter, thus reducing labor and machine cost.
- Quality assurance increases because there are fewer parts to inspect.
- The chance of parts not fitting an assembly decreases, reducing the rejection rate of a product.
- Less material results in a lower cost of the product.

In addition, with fewer parts, the product is more functionally reliable. With fewer parts that might break or malfunction, maintenance and support costs of the product to the consumer decrease. A single plastic part may perform numerous functions that would require many metal parts to perform the same functions. The plastic part may have structural, cosmetic, and flexible features all designed into it. This reduces the cost of design because the design requirements are being solved concurrently, not serially. The product savings come in the reduced development time for the design of the product.

Matching Design and Plastics Technology

This section defines the criteria for material selection for strength, deflection, temperature, and most other physical properties of materials used in plastics design. The plastics designer must select the correct plastic for the final product's intended use. The designer does not want to use very expensive plastics to perform tasks that do not require high durability when a less expensive plastic will suffice.

Strong, Hard, and Tough Materials

A strong plastic material is defined as a material that can withstand shock and abuse without breaking. Hard materials are scratch resistant. For example, a handheld drill requires a plastic that can withstand drops and that can withstand the vibration of the electric drill motor and gears. Polycarbonate is an example of a strong plastic material used at or near room temperature.

Fillers are materials that added to plastics produce a stronger and tougher plastic over a wider temperature range. Fillers have different physical composition characteristics than plastics. They can make a plastic stronger, but perhaps more brittle, and can even make a plastic electrically conductive.

Glass fibers are an example of filler material. The fibers could be 2 mm long and 1 micron in diameter. A percentage of the volume of the filler material, 5 to 30 percent, is mixed with the plastic material. Because glass is strong in tension, the addition of glass fibers to the plastic makes it a stronger, albeit more brittle, material. By adding glass fibers to a plastic, parts produced from the plastic are stronger and lighter than plastics produced through conventional injection molding methods that do not add fibers.

Plastics can also be made stronger by adding fillers to the plastic. Glass fibers have a very high tensile strength. When glass fibers are added to a plastic, such as a polycarbonate plastic, the resulting plastic becomes much stronger in tension. The new plastic has lost a percentage of its flexibility of polycarbonate but it now has the added strength of the glass filler. The Pro/ENGINEER designer must consult plastics companies and plastics data books to determine which combination of plastic and/or filler is best for their design for high-strength applications.

Deflection

Plastics can be chosen for numerous types of deflection purposes. The deflection might be repetitive, as in a living hinge, or a one-time event, such as with a snap feature. For snap features that require only a few deflections during the life of the product, a plastic such as acrylinitile butadiene styrene (ABS) might be selected, depending on the degree of deflection. This plastic is well suited to returning to its original position after being deflected. However, after about seven years, this particular plastic does start to lose its original memory (at room temperature) due to stress relaxation and begins to not return to its original shape completely.

The point of the ABS example is that the designer must be aware of the intended life of a product and select a plastic that will meet the requirements. The designer must also be aware of the maximum deflection a given material can take before the plastic passes its plastic yield point (the point at which it does not retain its original shape or perform its intended functions).

Temperature

Temperature affects each type of plastic in a different manner. Because thermoplastics are made from pellets that were once solid and then melted to form parts, the parts can also melt. These plastics have an upper temperature above which they will not be effective in performing their functions. As the temperature rises, a plastic may begin to change color, and eventually lose rigidity and become weak and soft. At cold temperatures, plastics tend to become more brittle and more easily fracture upon impact.

Other Physical Properties

The designer must select the correct plastic for the product requirements. Each plastic is different. By adding fillers to a plastic, you change the physical properties of the plastic. By combining plastics and other materials (such as fillers), plastics perform in various ways. Many thousands of plastic materials are available. Obviously, the combinations of plastics and other materials are a factor greater than many thousand. This type of information is available to the designer via plastics data books, which are normally geared toward helping the user select the preferred plastic.

Summary

You should have a basic understanding from this chapter of why and where plastics may be used. Parts are designed with plastics as the material of choice for appearance, integrity, and cost savings. The next step is to apply the material understanding to a design philosophy, the topic of the next chapter.

Chapter 3

Design Philosophy

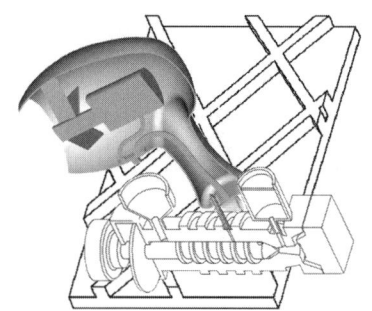

Initial Analysis and Project Planning

Introduction

Pro/ENGINEER is a modeling tool relatively easy to use to create complex plastic part and product designs. Experienced plastic designers are often pleased with the amount of geometry Pro/ENGINEER can accurately create in a short period. This chapter addresses issues related to starting a plastics design, with an emphasis placed on methodology. The discussion addresses issues such as selecting design manufacturing technologies and applying the properties to your project planning.

In using Pro/ENGINEER for plastics design work, the user must be aware that the program's feature count for finished plastic components can become considerable when all of the finishing manufacturing drafts and rounds are applied to models. This chapter covers design approaches that avoid creating geometry difficult to modify.

The experienced Pro/ENGINEER user plans ahead for splitting product design models into components for finished detail design where this is necessary. This chapter also covers a number of techniques that help identify component piece parts early in the product design cycle, thereby avoiding unnecessary work and rework.

Initial Design Planning

Thinking about a project in its finished form helps the user visualize how the product would be constructed in Pro/ENGINEER. Consideration given to the fit and form of a design is as important to the end user as it is to overall profitability of the product. Therefore, it is important to consider the user early in the design cycle. The following section discusses this important topic.

Considering the End User

Designers are faced with bringing concepts to reality, and they must deal with the fact that the detail of an initial product concept can vary widely. Some projects are taken to soft model form by an entity such as an industrial design department, whereas others might be left for the product design department to conceive. Designers may also find themselves as part of a multidisciplinary team to which they bring their product and part design knowledge.

Decisions at the design planning stage involve such considerations as producing a product that is designed for both manufacture and the needs (form and fit) of the end user. For example, what good would an easily and cost-effectively manufacturable video game controller be if it weighed 5 kilograms and could not fit in your hand? The design of a product must balance the needs of the end user with such considerations as ease and cost of manufacture. Project planning might also involve issues concerning the safety and security of the end user while using a product.

Known specifications (product specifications or design limitations) are always a design consideration, and knowledge of other factors affecting the desired final product help to inform the designer about what types of features they are likely to be modeling. For example, a minimum wall thickness or an assembly technique (e.g., screws holding pieces together) for the final product might be identified at the product conception stage. The following is a partial list of some of the issues that may need to be addressed at the outset of a product design to ensure a successful outcome. Your company may already have a similar list it follows when working on a project plan or specification for a product.

Initial Design Planning

- Must be able to be assembled by the consumer using conventional household tools.

- Must be functional after being dropped three times from a height of 6 feet.

- Must be flame retardant. Consideration needs to be given to the type of material and wall thickness.

- The volume of sales predicted is extremely high. This could result in a specific assembly requirement for high-volume mass production. The following illustration shows a product modeled for concept development.

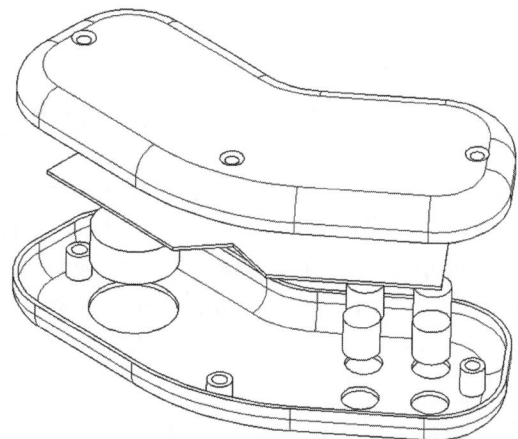

Rough assembly diagram showing component assembly order, modeled for concept development.

- Must be recyclable. There must be a way to disassemble the product.

- The product target weight is specified. This may affect the material selection.

- The end user must not be able to disassemble the product. You may need to use tamper-proof screws or ultrasonic welding as the holding mechanism.

- The product cosmetic finish is specified. This will lead to a draft specification for the chosen texture.

- Must fit comfortably in a shirt pocket.

The foregoing list of possible product specifications would certainly have an impact on the approach you would take to designing plastic parts or products. All are a reflection of the end use of the product or of the end user's safety and security.

Technology Limitations

Designers must often consider the limitations of the manufacturing technology selected to produce a particular part. Pro/ENGINEER will allow you to model a component or product that is unmanufacturable. It is up to the designer to consider the limitations of the manufacturing and material technology as early in the product design cycle as possible. This reduces rework and expensive delays in the project schedule.

Plastic resin suppliers, part suppliers, and mold fabricators should be consulted early in a project cycle if you are at all unsure you are designing within the limitations of the manufacturing technology chosen. Injection molding a bathtub, though it would be possible, would be a needlessly expensive project to implement when there are other, less expensive, technologies that suffice. Consider some of the limitations that could exist in the following list of physical testing to determine how the limitations would affect your design. The illustration that follows the list shows disks indicating maximum diameter reference to the gate location.

- Product must meet the specifications of a flame-retardant test. This situation will force you to design with a specific wall thickness and material to meet the criteria of the test.
- Drop test requirements could steer you to specific materials and design techniques to strengthen the overall design.
- Material specifications specific to an industry such as food-grade plastics might restrict some of the design criteria, such as minimum allowable radius for cleaning the product.
- Material flow (rheology) rates could restrict or complicate the design. A design might call for a material that has a minimum flow distance of 150 mm before freezing off (no longer flowing) in a mold.

Visualizing Product Designs 59

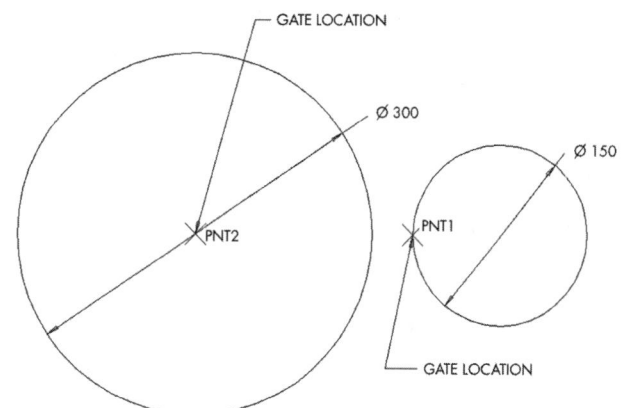

Disks showing maximum diameter reference to the gate location.

The illustration shows two disks. The disk at left shows a gate in its center, whereas the disk at right is gated on its end. Both disks show the maximum diameter reference to the flow distance under normal molding conditions for the chosen material.

The previous partial list of technology limitations will give you a sense of how this subject influences the overall component design in a product. However, end use and technology limitations are only two factors affecting design approach. The sections that follow explore others considerations.

Visualizing Product Designs

This section focuses on further identifying product design requirements to visualize what Pro/ENGINEER approach would best suit a given project. There are basically two approaches to product design. One is a "clean sheet" (top down) approach in which the designer has the freedom to model whatever is necessary to complete the design. The second approach starts with design criteria established (bottom up) before design modeling begins, which often involves reusing existing components and working within established guidelines.

To demonstrate the nature of these two approaches, the following sections present design approaches (top down and bottom up) taken to a common object (a video game controller). The differences between the two methods and how they affect the modeling task become clear as the details of each are expanded on. The following illustration shows the finished video game controller.

Finished video game controller assembly.

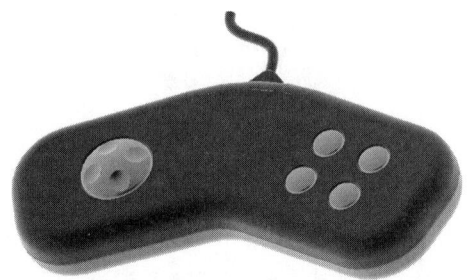

Top-down Design Approach

Pro/ENGINEER offers users a number of tools to manage designs and design data within assemblies. An assembly may be created that has components defined but not placed. In other words, this functionality allows users to keep a bill of materials up to date whether or not a part or product design has been started.

Starting a product design under limited existing restrictions (that is, minimally bottom up) allows the designer to take the approach of first developing the overall shape of a product. Rarely is the designer unable to identify design approach restrictions from whatever design criteria exist. In the case of modeling a video controller, the designer would begin by examining known (existing) product criteria (what the product concept is trying to achieve) and identifying the design restrictions indicated by the criteria.

Identifiable Restrictions

Examining and assessing design requirements enables the designer to establish a starting point from which to begin modeling. In the case of the video game controller example, assume that the overall cosmetic shape of the product has been identified, and that a list of product features is available, perhaps in a formal product specification. The following items, representative of such a list, pertain to the video game controller.

- A cable is required, with a known connector type. This can be an off-the-shelf item.
- The product must not be a burden to hold or use. Weight and form (shape) is roughly defined.
- The product is intended to be used with two hands.
- There are four buttons on the right-hand side.
- There is a four-position rocker switch on the left side.
- There will be a circuit board.
- The marketing department is looking for as thin a profile as possible.
- There will be a textured finish on the outer surfaces.
- The anticipated volume of sales is high. The assembly technique must be investigated to determine if the product should be designed for robotic assembly.

One approach to working on the layout for this product design would be to create the geometry and keep up to date with the component design as the process moves along. This approach employs a concurrent design methodology. That is, as design issues are identified, they are dealt with as they appear. Looking at a series of possible development steps for this design will show one method of approaching the product design task.

EXERCISE

Phase 1: Model Exterior Shape

Start a product design by modeling the overall size and shape of the product. Make the design modifiable about the parting datum plane to allow changes to the overall thickness of the product. The concept used here will be to initially model the product exterior shape, giving freedom to the overall design. With Pro/ENGINEER, the product model will be dimensionally modifiable once the assembly and component restrictions are taken into account. The following illustration shows the initial model shape of the product.

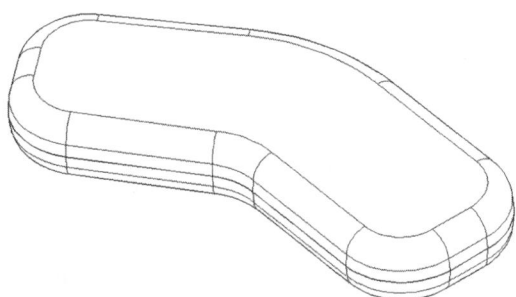

Initial model shape of the product.

The previous illustration indicates that the initial product design reflects the marketing department's request for a thin product design. For this reason, draft is included in the side walls. Consideration has been given to the manufacturability of the design in that the parting plane has been roughly selected. Whether or not the design is functional once the internal components are added is still to be discovered. A flat datum plane has been chosen as the parting plane for the design. In the final product design, the parting surface might not be a single, flat plane.

Phase 2: Initial Component Layout Assessment

With the overall product shape defined, the designer can move on to the layout of the internal components to determine if the product shape is usable in its current form. A quick, rough-modeling approach for the internal layout and tallest features will suffice for making this decision. This is a rough check to make sure that the internal components will fit into the overall shape. To perform this type of analysis, the designer has the following options available in Pro/ENGINEER. The illustration that follows the list shows the rough model.

Visualizing Product Designs

- Pro/SURFACE could be used to quickly model the internal components in surface form without creating another part file. Modeling components in this fashion would be suitable to try out the overall space claims of the internal structure to make sure all of the components will fit and function in the product.

- A skeleton model could be created, showing the space claims in surface structure, and placed on the product shape as an assembly. Using this technique (discussed later in this chapter) will allow information to be transferred from the skeleton model to the individual piece parts when they are created later in the product design process.

- Individual piece parts may be roughly created and placed on the product shape model as an assembly. This practice is common in concurrent design in which a number of people may be working on detail design at the same time. The freedom to have more than one person active in designing the product and its components should reduce the overall design lead-time.

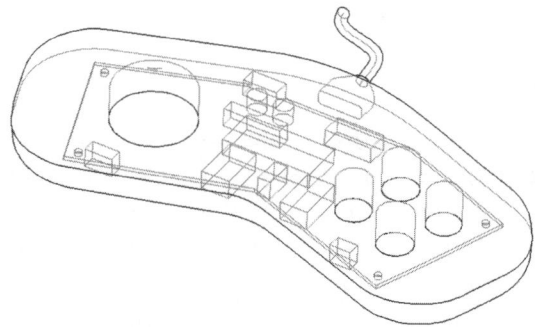

Components roughly modeled and placed on the product shape model (tangent edge lines not displayed).

➥ **NOTE:** *The techniques in the previous list are described in detail in Chapter 8.*

With the internal components roughly modeled, the designer is able to look at the overall assembly and make design decisions. The ability to quickly reposition components and modify shape dimensions allows the designer to tweak or fine-tune the component layout and the overall product shape. If major changes are required to the model's shape or its components, the time investment up to this point is minimal. The following are design considerations that might be assessed at this juncture in the process.

- Is the product shape model of suitable size?
- Are the components reasonably placed? That is, do they lend themselves to good ergonomics and functionality? The spacing of the buttons, switch travel, and so on might be considered in this regard.
- Is there room for the cable assembly to attach to the circuit board?
- Where will the fastening features need to be located? Will screws, snaps, or an alternate form of fastener be used for the assembly?
- Is there room for company labels and logos?

By employing the technique of "component accrual and assessment," the designer can make changes quickly, because the component models involved in the preliminary design have reasonably low feature counts. Few dependencies are formed, which means that there are fewer to deal with when modifying features and changing the location of components.

In the case of the video game controller, the product shape itself does not yet have shelled geometry or holes for buttons. Once the components are positioned, a number of people may be assigned to finish the component design with little threat to losing the design intent as the details are developed. The parting plane or surface for the product shape can be placed in its final position. Subsequently, the product shape may be split into the required number of components and the detail design of the components started.

Bottom-up Design Approach

Designers are often asked to consider using existing product components in their new design, which suggests employing a bottom-up approach to the design. This procedure is common among manufacturers who have existing designs that work well. Based on volume purchasing, extending the use of an existing component can lower the cost of design and manufacture, warehousing, and inventory control. In addition, tooling costs are avoided if off-the-shelf components are selected for use in the product design.

In this approach, existing components might be combined as a layout or assembly to serve as a staring point for product design. The design might include a couple of pre-selected components or include pre-selected subassemblies.

Laying out a new design with existing components can also serve as an exercise to see if it is possible to use existing items. You might discover that you have freedom in some areas of component reuse and restrictions in others. For example, in the case of the video game controller you might discover that you can use existing cable, buttons, and switches but need to redesign the circuit board to fit a new product shape. The following sections describe the major phases of this type of bottom-up approach.

EXERCISE

Phase 1: Gathering and Assembling Known Components

Collecting existing models and/or modeling existing components is the starting point for this approach. It is not necessary at this stage to model all components in complete detail. You are attempting to model the functional parts of all components accurately toward assessing the workability of the overall design and assembly. At this stage you are considering such things as known clearances between assembly components and clearance for mechanisms that move as you model. The finished product of this exercise will be a subassembly, shown in the following illustration, used as a starting point for your overall product shape model.

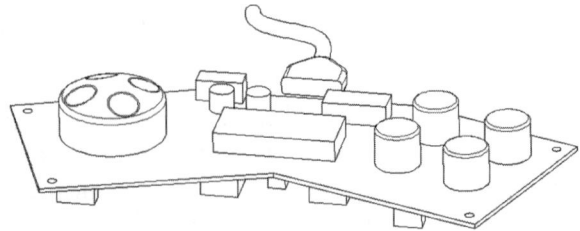

Subassembly showing existing components roughly modeled.

With the design in its current "flexible" state, it is possible to develop the shape of the overall product around existing geometry. Some restrictions exist that may cause the designer to sacrifice some of the design freedom. The following list identifies some of the freedoms lost in the case of the video game controller by employing this technique.

- The existing circuit board has already established the location of the switch and cable components.

- The overall height of the product is partially controlled by the existing height stack-up of components.

- The product shape needs to be large enough to enclose all of the components in their fixed positions.

Phase 2: Laying Out the Product's Outer Shape

At this stage, the model representing the outer shape can be started and assembled onto the subassembly of components. The dimensions of the model will need to be adjusted to suit the space claim requirements of the subassembly. The designer still needs to establish overall fit and form for the exterior geometry, as well as define the parting plane or surface that will be used to split the outer shape into piece parts. Once the outer shape is roughly defined, the outer shape model part may be divided into components for detail design, and the design may continue under the dimensional restrictions of the subassembly.

Video Game Controller Wrap-up

The two approaches (top down and bottom up) previously described represent fundamental methodologies that may be used with product designs in Pro/ENGINEER. Most projects will be best represented as a combination of the two approaches. It is common to have existing components used in a design when applicable. It is also common practice to design components specific for the application. Therefore, it is also common that both approaches have their place within a single design project or alternatively used for the numerous models of a design project program.

Identifying Components in Product Design

Pro/ENGINEER, being a solid modeling tool, allows the user to quickly visualize geometry created on-screen. Changes to geometry are redisplayed rapidly, with shaded images that allow you to visually accurately verify the results of design modifications, eliminating in most cases the need to have a physical model produced. Experienced users are more comfortable with determining the validity of what is on the screen than are beginners, but with time in the trenches and skilled use of the analysis and display tools, users normally develop this talent.

Creating plastics designs in Pro/ENGINEER is not as intuitive as a building-block type of design in which components may be machined rather than mass-produced, and for which the requirement of draft does not exist. In the building-block technique, you model components independently and assemble them as it suits your need. The design shown in the following illustration would fall into the building-block category.

Example of a simple building-block design layout of components.

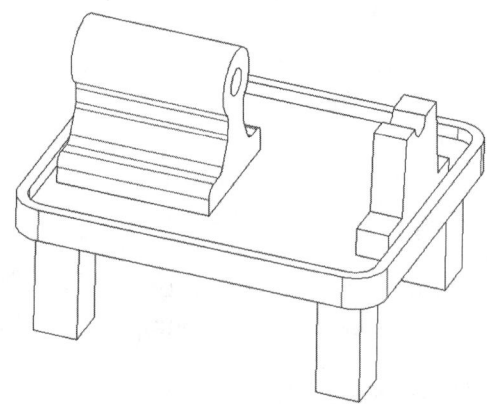

Plastics product design usually requires that parts be mated, which supposes that there is some sort of feature that defines the location of mating. Generally, the parting plane or parting surface, as the case may be, serves as the platform from which components are modeled. However, there are exceptions to this rule, based on the type of design you are creating and the technology to be used to manufacture the design. The design technique addressed in this section is related to multiple component designs in which components share at least some attributes, such as internal components, outer profiles, or switch buttons located on the shared parting line of housings.

In this type of design it would be unusual to be able to quickly model a product, add draft to it, and have it manufactured without first giving consideration to the overall manufacturing process and design requirements. When dealing with designs of this nature, it is sometimes necessary to model the design a couple of ways before you can come up with the best solution to the design problem. This is demonstrated by the electric razor assembly in the following illustration. A quick model of the overall shape created without draft and detail radii is made to help identify how the product will be manufactured.

Identifying Components in Product Design

Electric razor outer shape.

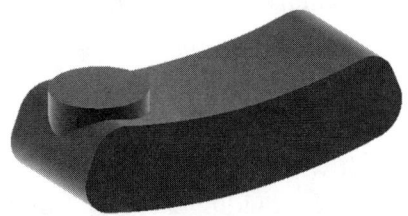

This is a simple design used to test the process of splitting the product design into components. In this case, the product logically divides into fundamentally two components: a top and bottom. Assuming the requirement is to injection mold the components of this design, you would have three basic ways of separating a design of this nature into two components. The following illustration shows the design split into top and bottom components.

Design split into top and bottom components.

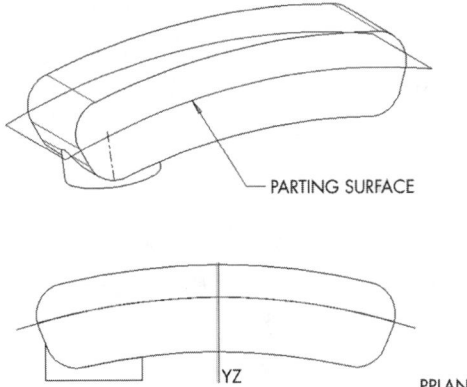

Splitting a Design into Top and Bottom Components

When you split a design into components, using the horizontal datum plane as the reference parting plane, you are able to more easily view and assess component models, taking into consideration the manufacturing process used to build the parts. In this case, the parting surface is curved. Making a list, such as the following, of the manufacturing and end user issues will help to uncover the pros and cons for creating the design in this manner.

- When draft is added to the side walls of the components, there will be a line formed (parting line) representing the joining point of the two components when they are assembled. The line represents the position in the model where the product shape will be split into components.

- It will be difficult to achieve the opening for the cord on the housing without providing a slide in the mold.

- It is necessary to have an irregular parting surface to maintain a constant wall thickness.

- The assembly technique used to assemble the internal components is not straightforward because the cord has to be routed through the lower housing.

Splitting Components into Front and Back Piece Parts

When determining how to divide a product design into individual piece parts, it is important to note that the parting surface need not be a flat, planar surface. In the example of the electric razor, you would offset the parting surface and attempt to determine where the piece parts will separate. The assembly task using this split example would be equally as complicated finding the parting surface. Consideration to the final assembly task of the product needs to be taken into consideration when selecting the parting surface location, as well as the manufacturablility of the piece part. This design would need to be split into more than two components to be manufacturable using plastic injection molding technology.

The resultant curved outer walls of the components would make the manufacturability of the design questionable. This is because undercuts created by the curvature of the part walls might prevent employing a simple mold separation process once the shape were shelled to obtain a constant wall thickness. It would require difficult and expensive tooling to allow for separation of the mold when an alternative method of creating the components might be used. A design such as this might require that you further split parts from the outer shape to allow for the realities of the manufacturing process, as indicated in the following illustration.

Identifying Components in Product Design

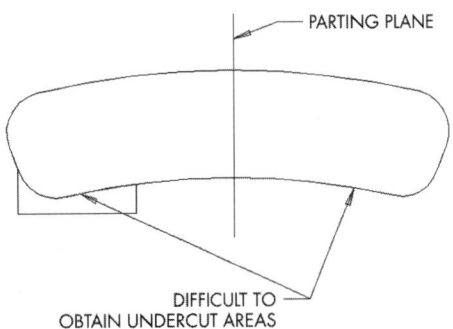

Design split into front and back piece parts not possible in two parts.

Sometimes product designs have to be divided into more than two external components. Try to select a methodology that minimizes the total part count while preserving the intended outcome. However, keep in mind other issues when reducing part count. For example, if a design required batteries replaceable by the consumer, it would not be advantageous to require the consumer to disassemble the unit to get at the batteries. Omitting a battery access door would be counterproductive in this case.

Splitting Components into Left and Right Piece Parts

When a product assembly is designed to be parted about the center datum plane, ease of assembly is enhanced. However, the trade-off in this case is that there is now a joint all around the product in the center. This situation may or may not be acceptable to your design criteria (such as for aesthetic reasons), and you may be forced to use more than two components to manage the location of the parting line. The assembly of the internal components is relatively straightforward in the design shown in the following illustration because everything can be assembled from one side.

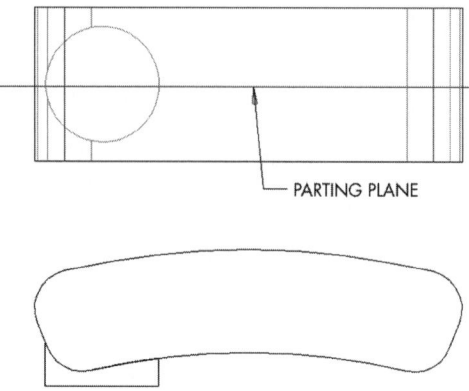

Design divided into left and right parts.

Electric Razor Wrap-up

The task of splitting an outer shape design into components is relatively easy with Pro/ENGINEER because the program allows you to try a number of ways of achieving the desired result. Pro/ENGINEER's analysis tools help you check for undercuts or non-manufacturable areas in your piece parts without a major time investment. Once the components are identified, the detail design of the components may begin. You would start by adding the manufacturing draft and external rounds on the piece parts.

> **NOTE:** Pro/ENGINEER analysis tools are discussed in Chapter 17.

For designs whose parting surfaces are not planar, the designer needs to use curve-driven draft to maintain the integrity of the common profile (i.e., where the components join). It is important to note that you will need to divide some designs into a number of piece parts for the overall design to be producible and the product manufacturable.

Using Pro/ENGINEER, the designer has reasonable assurance that the time investment in the design planning stage will be minimal. This is because the designer can in the early stages of the design cycle quickly model and assess the various product design options and likely component splits. The Pro/ENGI-NEER feature count is relatively low at this point in the design cycle, and changes required to the overall shape of the product to enable manufacture of components can be quickly attained. However, as with any modeling package, once manufacturing draft and rounds are placed in a design, it will be more difficult to rapidly make changes to the design.

Product Design Modeling Techniques

It is important to note that all product and product component designs used in this book are started from a generic start part. The start part includes the default datum planes, coordinate system, and default view names for common viewing directions. In Pro/ENGINEER, the accuracy is also preset internally in the start part to be within the range of .0003 to .0008 because of the complex nature of some of the fillet radii surface requirements. Geometry checks are not permitted in the model data, unless they can be explained as part of the design intent. Tiny edge geometry check errors may be permitted if you do in fact want a tiny edge in your design. Misalignments and unattached features are not permitted in a quality component file.

There are a number of techniques associated with managing plastics design in Pro/ENGINEER. When product designs involve a number of components cut from the same original master part file, there are a number of ways to handle the design task. Designers should look closely at their design job and try to use the technique that best suits a given application.

Pro/ENGINEER assemblies are particularly important to the process of designing mating parts. Therefore, the designer has a readily available design sanity check at any point in the process simply by pulling up the assembly. Experienced users might actually work in assembly mode. However, the risk associated with this practice is that they may cause components to become dependant on one another during the regeneration of the model.

> ↦ **NOTE:** *Chapter 4 addresses the assembly of components under Pro/ENGINEER and restrictions associated with component assembly.*

In a product design job that requires positive matching of surface structure and shape between components, it is usually best to model the components as if they were one part and split the piece parts from the original design. Sometimes you do not want to carry the overhead of the complete model inside each component model and want to capture what is strategically useful to you in the component design, transferring this information only to your Pro/ENGINEER models. Skeleton models allow the selective capture of such data.

Datum curves, surfaces, and so on may be created on a skeleton model and shared with components by using a skeleton model in your product assembly. This technique saves you from having to repeat the core data for a design more than once, and the design may be at least partially managed from the skeleton model. One other feature of skeleton models is that they may be used as kinematics tools to, for example, move linkages or open doors, depending on how the model was created and how the piece parts are assembled to the skeleton model.

Curling Iron Skeleton Model

Skeleton modeling is a subset of the Pro/ENGINEER assembly module. A skeleton part file may be created within an assembly's model tree to use as a mechanism to hold information references, by component, within the assembly. A typical skeleton part might contain reference geometry consisting of datum curves, surfaces, and points. Information from the skeleton model may be used as a visual aid, or to create geometry within the various piece parts of an assembly. An example of a visual representation would be the geometry for a camera battery for which the designer wants to make sure the clearance within the battery housing accommodates the battery size.

The illustration that follows shows a Pro/ENGINEER skeleton model of a curling iron. The skeleton model has a number of reference features in it to aid in the overall modeling job for the plastic housing components. The features modeled into the skeleton model serve different purposes and therefore represent attributes of the overall design. This skeleton contains the features described in the list that follows. These features are common to the product housing design task.

Product Design Modeling Techniques

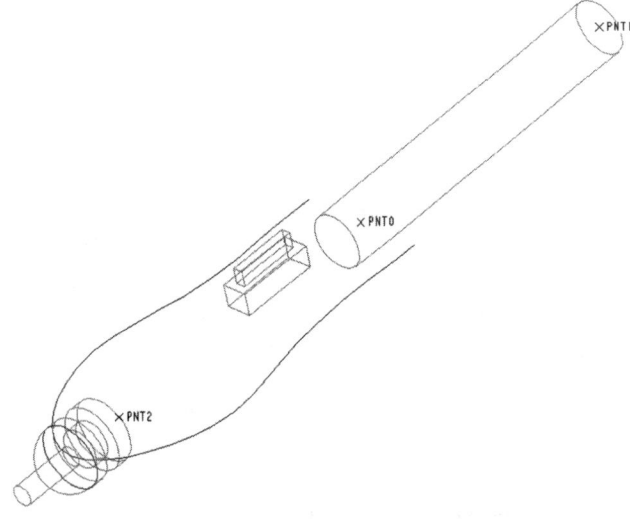

Curling iron skeleton model.

- Three datum points to represent length positions.
- A datum curve representing the outer contour of the plastic housings at the parting plane.
- A surface model representing the maximum outside diameter of the metal shank of the product used as a space claim device for modeling the plastic parts.
- Another space claim model of the switch assembly to ensure that the housing encloses the switch.
- The cord grommet is also modeled as a surface model to aid in the housing design.

The plastic housing design is made easier when common items are shared from a skeleton model during the design process. By applying the references in the skeleton model to both the top and bottom housings, the designer can be assured that they are in fact the same and will mate properly. The illustration that follows shows the initial housing design, which references the skeleton model.

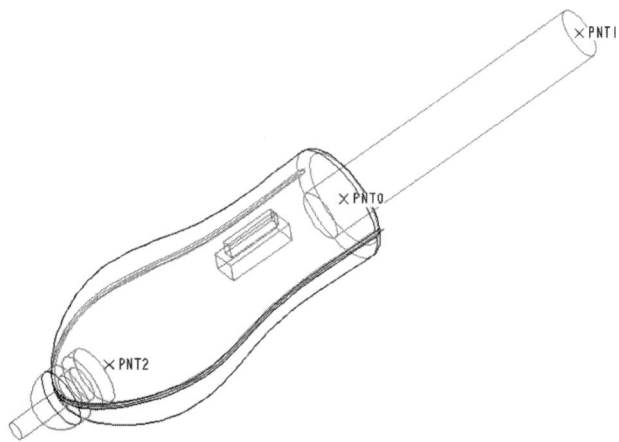

Curling iron initial housing design referencing the skeleton model.

There are other Pro/ENGINEER options when users are interested in sharing data across components. You can assume that some product designs are best represented by a combination of techniques for managing the overall product design. There are two basic approaches to taking a product design consisting of a number of components designed simultaneously and splitting the design up into individual components. The sections that follow discuss these approaches.

Product Individual Piece Part Creation

One technique would be to design the external geometry of a product, create enough copies of it to allow for independent components, and create your piece parts from the individual files. This technique is common practice among designers who want to keep their components independent from one another.

The model is initially a product model representing a number of components. The necessary parting surfaces are added to the product design prior to creating individual part files. Once the individual part files are created, future changes to the external geometry are duplicated in the other parts to maintain a common core part.

> **NOTE:** *Chapter 4 discusses modeling tricks that help users identify and keep track of common features in piece parts. To touch on the subject for this discussion, designers should identify what these root features are*

Product Design Modeling Techniques

and ensure that they are in the model prior to splitting the product into components. The illustration that follows shows an exploded assembly of an external housing design.

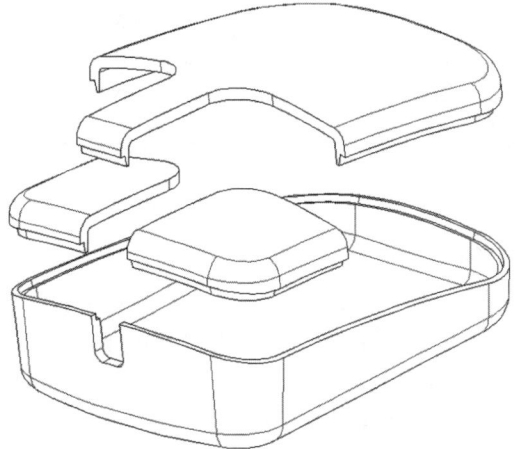

Exploded assembly of a computer mouse external housing design.

Mating coordinate systems of the individual components can aid in the assembly task for this type of product design. Part modeling from a product model makes for a trouble-free way of ensuring your component design has not altered the overall intent of the product. Another approach, similar to this, retains the model dimensions in a common master part file. This is discussed in the section that follows.

Product Design Using a Master Part File

The example in the previous section will be reused as a comparison when designing with a master model file. Consider that the product design was complete for the computer mouse exterior housing shape in the previous example. Pro/ENGINEER allows the user to create piece parts using the original part as the master part file. The technique for this procedure differs from the previous example in that the original design file may be modified in size and the individual components will all update upon regeneration.

The important feature of using this technique is that your individual parts are dependent on the master part file. Some company standards for component filing systems will not allow this procedure because components are expected to be independent in case they are also used in other future designs.

Consider the example of a plastic house modeled using a master part file where your goal is to separate the components for easy shipping in cartons. The product is intended to be assembled by the purchaser. The basic design is subject to constant churn while you are working on it, as the marketing department has not finalized the requirements. The following are parameters of which you are aware.

- The style and preliminary size of the house.
- The wall thickness of the plastic parts
- Assembly by the purchaser without the use of tools.

Given the above criteria, you may model the house, attempting to stay as flexible as possible with the overall dimensional properties. The object is to define the overall shape and start on the component design required to assemble the unit even before marketing has given you the final dimensions. The toy house to be designed is shown in the following illustration.

> **NOTE:** *When concurrent engineering is a requirement between the manufacturing site and the design site, the master modeling procedure can reduce product development lead-time considerably if the manufacturing site is also a user of Pro/ENGINEER. When the model is updated, the manufacturing files may update as well as the design files.*

Toy house design, release 1.

Product Design Modeling Techniques

The technique chosen for this modeling task is the master part technique. The process for creating the piece parts is relatively straightforward. Parts are created for each individual component. These parts may be typical start parts with only datum planes and coordinate systems. The following list of steps may be more intuitive.

1. Create the initial design file of the complete house design model including modeling the wall thickness of the plastic.
2. Create individual piece parts representing each of the components. These piece parts have no significant features in them. They are blank parts used only to reserve the component identity.
3. Create an assembly and assemble one of the component parts to it.
4. Assemble the original design file representing the complete house into the assembly.
5. Using Assembly Advanced Utilities, Merge Reference the design file onto the piece part. Make sure to use Reference and not Copy.
6. File the assembly and pull up the piece part file by itself as a part file.
7. Cut the component you are modeling out of the design file.
8. Repeat this process for the balance of the individual component part files.

You may then assembly the components again into another assembly to verify that the components have been cut out of the original design properly. Once this is accomplished, any change made to the original design will update throughout the piece part files because the original design file is only referenced in the part files. The front wall is shown in the following illustration.

Front wall modeled by cutting away the rest of the design.

This procedure can be difficult to manage in an environment where design files are shared between a number of users on different workstations. The components rely on the master file to be present and kept up to date. The updated file and front wall are shown in the following illustration.

Design file changed (left). House wall updates automatically (right).

Summary

This chapter has explored design philosophy and how it is applied to everyday design tasks. The material on designing for known specifications and early product design planning has shown the importance of thinking through the design process early in the cycle to avoid painful rework or restarts in the design. Now that you have covered preliminary design planning with an emphasis on discovering what it is you are modeling, the next chapter will carry on the discussion with focus on design planning and organization down to the piece part level.

Chapter 4

Design Planning and Organization

Planning at the Component Level

Introduction

The previous chapter covered overall product design techniques. This chapter discusses modeling plans and procedures for designing plastic components. Techniques learned in the last chapter for identifying components are covered at a part detail level in this chapter. The discussion also covers identifying root and buried features, and creating features as aids to modification of existing geometry.

Parting surfaces and split parting planes are covered in this chapter where they apply as root features in the overall design. The function and identification of root features will also be covered to unveil their powerful attributes further into the modeling cycle of a plastic component.

Planning ahead is important to the outcome of any design project. Modeling plans aid you in determining an approach to developing a design better managed for future modifications. To plan modeling well, you need to be aware of the attributes of plastics that play a major role in the definition of the overall model geometry at the earliest stages of the design. The following sections explore this issue.

Parting Surfaces and Parting Planes

The parting plane or surface in a design assembly is considered the main parting plane or surface between the core and cavity of the mold in an individual part. It is sometimes referred to as the shut-off between the core and cavity, although there could be one or more areas of sealing internal to the perimeter of the component.

In plastic component design, the parting surface where the core and cavity of the mold mate may be a shaped surface. Designers need not be restricted to considering parting surfaces as planar entities. For the purposes of this discussion, "parting surface" refers to the "condition," rather than to a plane, because a flat parting plane can also be considered a surface. The text refers to the "parting direction plane," which is the plane that defines the direction of pull between the core and cavity of the mold. An example of the definition will help to identify the difference between "parting surface" and "parting direction plane."

In the following illustration, a component is intended to act as a lid for another component with an irregular shape. The parting surface is curved, but the parting-direction plane is positioned such that the component may easily be ejected from the mold once manufactured. In this example, the manufacturing draft would be applied to the model referencing the parting surface. However, the draft angle would be measured from the parting plane to ensure the part ejects from the mold.

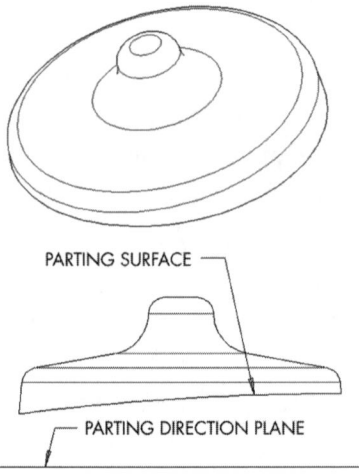

Parting surface and parting direction plane.

Parting Surfaces and Parting Planes

The previous illustration shows the basis of the distinction between the parting surface and parting direction plane in a relatively straightforward component design. There are also component designs in which the parting surface and parting direction plane are coincident. A good example of this situation would be a desktop penholder, shown in the following illustration, which simply rests on the desktop and therefore has a flat parting surface. It would also be most beneficial to the component and mold design to use this surface as the parting direction plane.

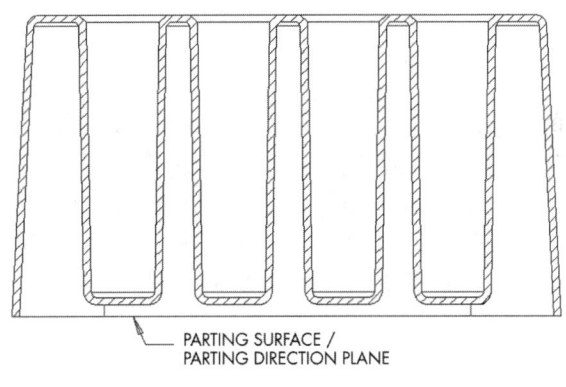

Flat parting surface evident on penholder.

Consideration is given in this chapter to the situation in which the mold parting surface is represented by a number of surfaces rather than a single surface. This situation is common when plastic components are designed to serve more than one purpose. For example, consider the component in the following illustration. The component is a top cover for an electronic enclosure. The enclosure has a front panel and rear panel identified as separate parts. There is also a bottom housing, which attaches to the top cover in the center of the product design. The parting surface on this component must start at the side of the part and travel around the front and rear panels.

Stereo housing parting surface.

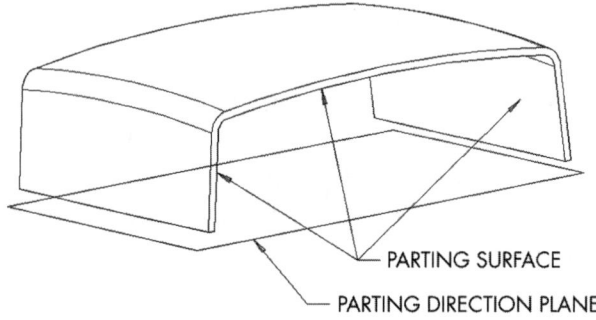

There are some instances where parting surfaces require special attention to define where the split between the core and cavity will take place. Consider the following example of a box with a tube. In order to mold the box, the parting surface would be a simple planar surface. When the tube is considered, it is no longer possible to have a single planar parting surface. The O-ring groove on the tube itself makes molding of the tube require a split along the center line of the tube. The hole in the tube will require a slide from the side of the mold.

When the parting surface is offset in this manner, the draft also needs to be considered during the modeling of the component. For example, the parting surfaces shown in the following illustration require that the external component draft feature be offset at the parting line of the tube. Pro/ENGINEER allows you to create draft with a split (that is, draft in more than one direction is applied to a surface or set of surfaces) on a single surface, making modeling of the component straightforward. The following illustration shows a cross-sectional view of the component. The draft has been exaggerated for display purposes.

Box and tube exhibiting split draft.

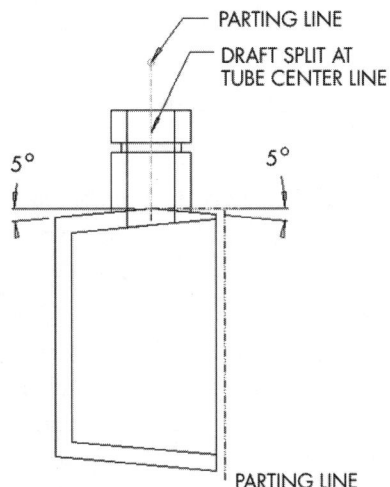

Identifying Root Features

Pro/ENGINEER may be considered a "history-based" modeling system. Not only does Pro/ENGINEER keep track of the features you create; it keeps track of the order in which you create them. Models can become very difficult to modify without such information. This is especially true if the user that created the model is not the same person modifying it. Depending on the availability and accuracy of this "historical" information, users find themselves either reaping the benefits of the parent/child relationships among geometry or suffering because of them. Model planning (discussed later in this chapter) can help you organize your design files in a manner that is reasonably identifiable.

When considering a design, you also need to give consideration to potential design changes at a later date. This approach is not uncommon to Pro/ENGINEER users because the software's ability to allow such modifications is usually a welcome advantage. Root features are central to the overall geometry of a component design and are most always found as parent features. Their placement in a model does not necessarily have to be at the beginning of the model tree.

✓ **TIP:** *Root features may be created where they make the most sense. They do not have to be at the beginning of a design file but may appear anywhere in the model tree. Recognition of root features can help you manage your component design, especially when dealing with mating parts.*

The housing shown in the following illustration is a simple example of a root feature. The first significant feature in the model is a square protrusion 50 mm across the flats. Before that protrusion was created, there could have been a start part with default datum planes, coordinate system, and so on. A start part is a good modeling practice, but is not a requirement of the software. The square protrusion will become the parent to the features that follow, known as children. The draft, outer round, and shell are children of the original feature. To modify the overall size of the object, you would simply modify the dimensions used to create the root feature.

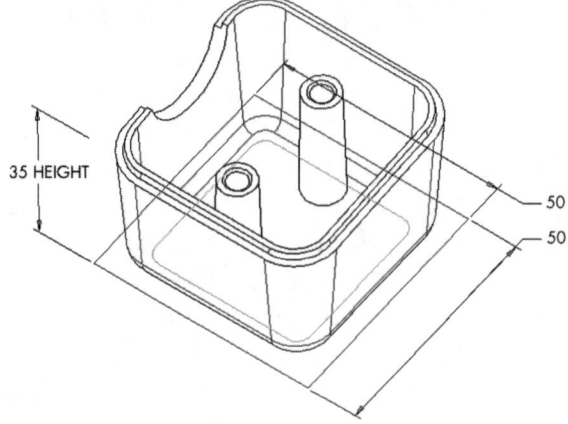

Square protrusion as root feature with its dimensions selected for modification.

For the model in the previous illustration, it would be reasonably easy to identify the root feature. The designer could use Query Select and pick through the added features until the protrusion highlights, or select the first protrusion from the model tree. Draft and rounds can bury (cover) original features, making it difficult to select them later in the modeling cycle. However, there are a few techniques you can use to make the identification and selection of root features easier, and therefore make modifications to the design more intuitive.

Visualizing Product Designs

Identifying Root Features

- Rename features as necessary to unique feature names easily recognized in the model tree.

- Use points as tags to act as controlling features in the design. Points could be used to identify a component height, the location of features, and so on. Points used this way are root features to some other part of the design.

- Datum curves may be used, which are easily identified in the model because they remain visible and may be easily selected.

- Other datum features such as planes and surfaces may also be used as root features in a design.

The following illustration shows the previous example using root features, which are easily identified to create the geometry. The feature count in this model is significantly higher because features have been added that serve only to identify other features. The datum curve has been added to allow access at any time to the external profile of the component. A datum plane has been added and renamed to show that it is used to control the overall height. If this model were 500 features in size, the root features that control the basic dimensions would be easily accessible to the designer should model changes be required.

Root feature datum curve easily accessed by the user.

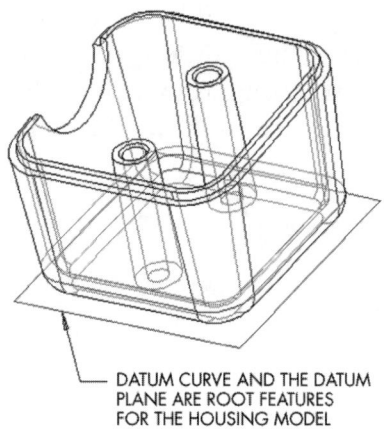

DATUM CURVE AND THE DATUM PLANE ARE ROOT FEATURES FOR THE HOUSING MODEL

The plastic housing in the following illustration is a more complex example of root features. The parting surface in this model is curved in two directions. Therefore, modeling this component with simple protrusions and cuts will be a more difficult task. The component also mates with another component, making it necessary to define a common profile at the parting surface between the two parts.

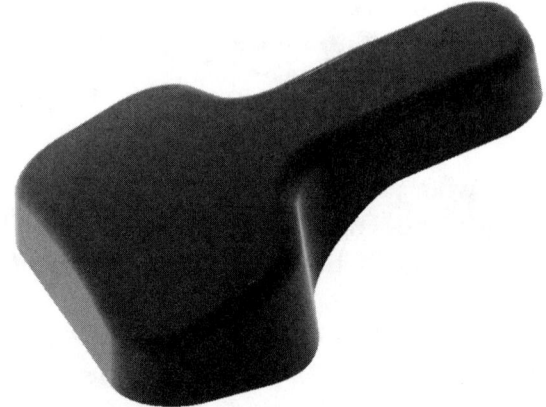

Complex curvature parting surface.

The following illustration shows the root features belonging to the previous illustration. The root features in this component are the parting surface and the profile common to both parts. When modeling the parts, the designer might start with a part containing the root features and copy it to another name to create the second part, or use a skeleton model in an assembly that contains the data for the root features. The illustration shows the component partially modeled.

Identifying Root Features

Root features of a model with complex surface and datum curves.

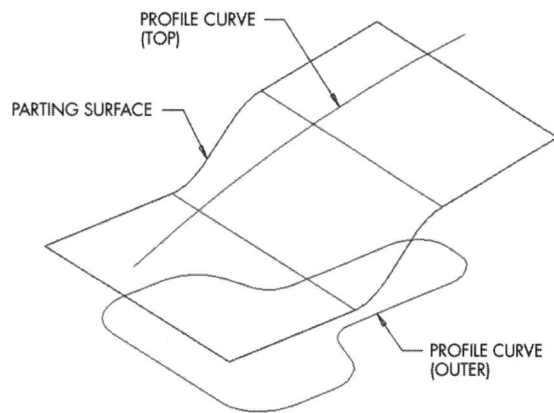

Looking at other common features of the mating components in the previous illustration identifies other root features that might be positioned in the model prior to making a second copy. If the screw boss locations were identified, the common position would be shared with the mating component. If there are other common profiles that might be shared between parts, these could be added in the form of datum curves, points, and so on to the original model.

Consider for the purpose of discussion that there are three screw bosses and two common profile cuts through the housings. Placing a datum point at the center could identify the screw boss locations, and the profiles could be identified using datum curve outlines. The following illustration represents what the root feature model would look like with all of the common features identified.

Root feature model continued.

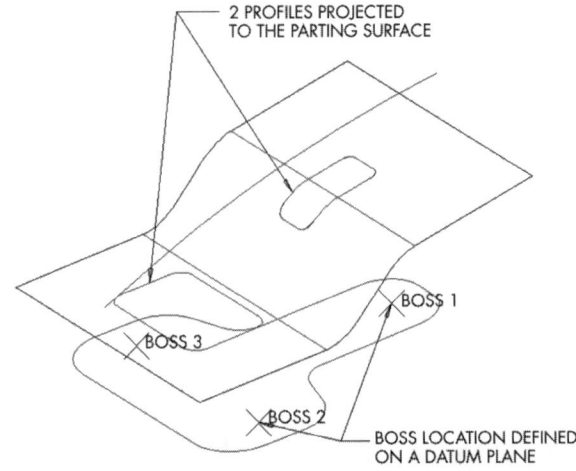

When initializing a design, if the designer is able to identify common shared features between components, the features may be modeled into a common part to serve as a start part for the individual components. This concept could be taken even further if the product you are designing is similar in many aspects to others. In this case what you are really designing is a product that could belong to a family of products. You therefore might be able to identify a common start component containing most of the basic features.

An example of the application of this approach would be for enclosures that vary only in height, length, and width, but that contain individual features, such as common cord entrance holes, button arrays, and so on. In this situation, you have an opportunity to create a functional root feature component that has all of the possibilities modeled as datum curves. You can then simply select which features apply to a given design.

Identification of Critical Specifications

It is unnecessary to say that the more difficult the geometry is the more difficult it is to describe. There is a growing tendency among companies racing to get their product on the market quickly to avoid or delay creating conventional documentation of design files (drawings). After all, there is a 3D model defining

Identification of Critical Specifications

the component. Why would you need a drawing to further define a model if you are using a toolmaker that works from 3D?

The answer is that tooling concerns for the designer might have more to do with individual features than to the overall form of a model. Perhaps there are dimensions critical to the proper functioning of the product. There might also be design features that could be misinterpreted by the toolmaker if they are not identified at the time of manufacture of the mold.

The reality is that in some cases you may not even know who will manufacture the tooling. Will the tooling even be fabricated on the same continent the factory is on? Will the tapered fit in your design be interpreted as toolmaker draft?

If you always use the same local toolmaker, you are able to drop by some afternoon and go over the design intent with them to help convey what the fit, form, and function of the components are. You might also be able to convey in the purchase order some of the key features you are looking for in a finished component, such as the overall flatness of the part.

It is always good practice to have the same toolmaker create the tooling for all mating parts. However, what if a mold has to be replaced for a component? It is difficult without creating an engineering drawing to go any further than making a few notes to remind everyone what is important to the design and filing the notes in the project file.

When there are critical requirements to be relayed to a toolmaker or the part inspection department, or simply to help capture design intent, a critical dimension drawing may be created. The critical dimension drawing seeks only to highlight what is important to the design. For example, it conveys which tolerances to take particular note of and what the inspection dimensions are intended to be.

Some detail could be highlighted in your 3D model if you are exporting your file to someone else who has Pro/ENGINEER. The probability is that some of the toolmakers you deal with do not have or are not going to get Pro/ENGINEER software. Therefore, you will be exporting a STEP or IGES file to the vendor. In these cases, and to make sure the design intent is passed on to the toolmaker and inspection department, it may be to your advantage to accompany the purchase order with a critical dimension drawing.

When identifying critical dimensions, remember that there is a specification that covers your generic model. The specification reflects standard mold making practices for plastic design. Consult your tooling vendor to find out what tolerances they work to when working from your 3D model. The following are items you may want to include in the critical dimension drawing.

- Geometric tolerancing for mass production of mating components
- The three geometric datums required for component inspection
- General material and overall notes reflecting the finished production part
- Dimensions on the mating features of the component (screw boss locations, external profile, and special tolerances such as +0.00/ -0.01)
- Texture requirements on the finished part
- Overall length, width, and height to aid creation of the manufacturing tooling
- Revisions to the component so that changes to the design may be traceable

In the following illustration of a critical dimension drawing, the component has to mate with another component around the external profile. There are screw bosses to be dimensioned, and manufacturing and inspection data has been added. The material block will simply say "Plastic" to maintain the generic nature of this book. It is noteworthy to say that when dimensioning the critical dimensions of a component feature, it is not required to dimension the entire feature. In the example of the screw boss, the only critical dimensions required would be location and height, which mate with the other component.

Critical dimension drawing.

The critical dimension drawing helps the designer convey design intent not only to the toolmaker but to inspection and manufacturing. The drawing also leaves a trail of design intent others may reference in the future. Now that you have an understanding of the role of root features and critical dimensions, you can begin to explore the feature creation process, covered in the following section.

Logical Feature Creation Process

"Logical" feature creation is to a degree subjective. What may be a logical process to one person may not seem logical at all to another. The discussion in this area focuses on the obvious and how to control it.

Pro/ENGINEER may be considered a feature-based modeler, which maintains a part history. If you should want to see how someone created a model, you simply replay the procedure using the Regen Info menu pick. This is a common procedure when designers are asked to work on a model someone else created.

Designers who took their Pro/ENGINEER training under previous releases of the software might have been told to always leave the draft and rounds to the end of the modeling process. However, this is not necessarily the best approach to a design and quite often ends up with the user fighting the software, trying to add rounds where they do not fit and draft on ribs that causes the finished rib to collide with something else. The reality is that if the model had been created with consideration given to the correct time for the addition of draft and rounds, the process of adding them would be more manageable.

It is quite common among Pro/ENGINEER users to model a component rather quickly and have the bulk of the model defined in its unmanufactabule form in less than 40 features. Manufacturable geometry means that the designer has described all attributes of a design in a 3D file. Therefore, a finished design includes all of the necessary draft and rounds needed to complete the design.

Logical Feature Creation Example: A Housing

Consider the following example of a housing created using logical feature creation. The housing has a wall thickness of 2 mm. It has two snaps at one end and two screw bosses and an engagement tab at the other. There are two ribs internally that divide the chamber into three sections, one section of which contains a lens window. To model this component, consideration has to be given to the mating component. The component profiles have to match when assembled. Solving this need becomes the starting point for the model. Looking at the design process, outlined in the following material, will help to reveal the logic behind the steps chosen.

1. Create the main protrusion using the profile of the design assembly as the external form. At this point, it is time to consider what can be modeled before the shell feature is added. If the external cosmetic features were added at this point, the internal portion of the model geometry would be created automatically during the shelling process.

2. Model the draft feature on the external walls and add the finishing rounds if they are a larger dimension value than the thickness of the shell. If the external radii are of a value

smaller than the thickness of the shell, the shell geometry is not likely to work when applied. The following illustration shows external features modeled before shelling.

External features modeled ahead of the shell (draft and radii).

3. The shell may be created at this time, resulting in a part with constant wall thickness.

The next modeling task is a choice of bosses, internal ribs, snaps, lens window, or the lip that interlocks the components. It might be argued that at this point it would be beneficial to model all of the geometry and add the draft later. However, you would then have to ask how stable the design was. What if the snaps had to be replaced by screw bosses? What if ribs were added or removed later in the design? How many support features are required to relay the design intent and/or completely model the required geometry?

It would be beneficial to model the features and their supporting features as a continuous string of features. This process follows sequential thinking. The modeling of the screw boss would be followed by modeling of the screw hole, draft, and any supporting rounds at the base of the boss. If you subsequently needed to remove or pattern the screw boss, it would be in a form that could be easily manipulated.

You might even consider modeling the ribs as independent entities if there is a chance ribs will be added or removed at a later date. To locate the screw bosses and snaps, datum points are placed as the first feature in the modeling sequence. The snap geometry should also be modeled in a sequential manner to allow for modifications later in the design process. The illustration that follows shows the finished model. The following are the feature counts required to complete the finished part design.

- External geometry up to the shell: 9 features
- Interlock lip: 1 feature
- Screw boss (one complete): 7 features
- Snap detail: 6 features

Finished component.

Model creation using this technique usually results in more features in your design than using another method. Not following the procedure may allow you to have a considerably smaller feature count model. However, you may not be able to change it without a major amount of rework.

It is also important when using this technique to mange features such that they do not tie the entire model together and create dependencies within the geometry. Strict attention should be paid when creating features to not dimension or align unrelated geometry to what you are currently modeling. The benefits of separating the design intent into identifiable design requirements will better enable you to manage your design.

Tips for Identifying Buried Features

When modeling plastic parts, Pro/ENGINEER users sometimes have to partially or completely cover over features used as root features in a design. This is sometimes referred to as burying features. An example of this would be the modeling of a screw boss from the previous section. The first feature in the modeling process for the screw boss was a protrusion created from a datum plane, the protrusion finishing where it intersects the existing geometry.

Once the supporting features are added to the original protrusion, it becomes difficult to select the protrusion for modification. This situation is common in modeling plastics because almost everything you model requires draft and finishing rounds. The following illustration indicates how the original protrusion might get buried.

Screw boss, with original protrusion buried.

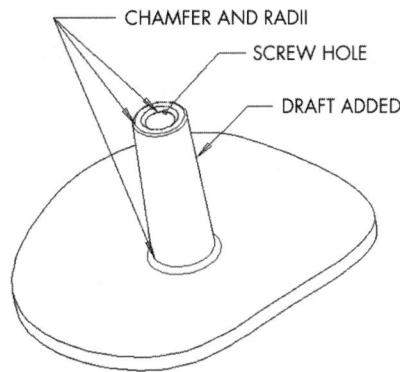

There are a couple of techniques that would aid the user in labeling or identifying buried features. One technique would be to rename the feature in the model tree to give it a unique name; for example, Screw Boss 1. Another technique would be to model an identifying feature prior to the protrusion. This technique would at least identify the feature before the protrusion, making it easy for the user to identify the screw boss protrusion.

If a datum point were modeled representing the center point of the screw boss prior to modeling the protrusion, the center of the protrusion could be aligned to the point. If the location of the screw boss had to be modified, the user would need only move the point, and all of the associated geometry would travel with it. The following illustration shows a partial list of the model tree identifying key root features.

Partial model tree list.

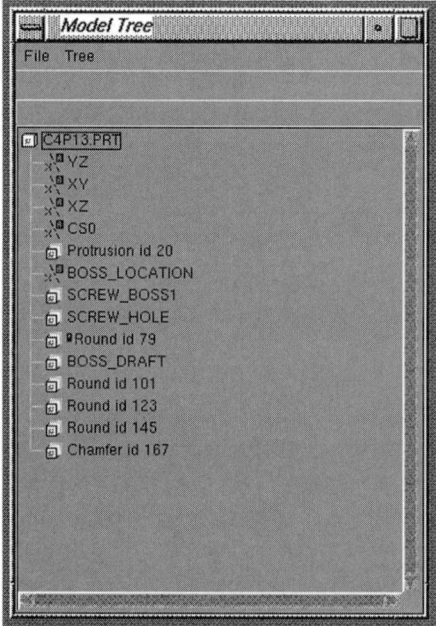

Creating a Model Plan

The previous sections have covered logical feature creation, identification of buried features, critical features, and identification of root features. This information may be used to analyze a design job and create a model plan, which is simply a reasonable technique or best approach to the modeling task. If your office has a number of Pro/ENGINEER users with varied experience, it may be worth taking some time to identify what you are creating, what order you would create it in, and how you would approach the design task, and to pass this information to your peers for review.

The information you supply should also try to predict what will most likely have to change during the modeling cycle to ensure that certain features are modeled with particular care. The modeling plan would be best developed away from the computer, avoiding the tendency to start modeling as soon as you start planning. The following are points to consider and questions to answer while planning your model.

- Is this a component or product design task?
- Does this part mate with other parts?
- What would my modeling approach be at the start?
- Should a protrusion be created?
- Should the geometry be swept?
- Should I start with a datum curve representing the profile?
- Does the shape have to be developed using surfaces?
- What features can be grouped together to enable easier modification of the geometry?
- Does any geometry have to be imported from a foreign CAD system?
- In what order should I create the geometric features?

This list could be greatly expanded. However, it is possible that the modeling task is similar to another component used at your site, which may give you the opportunity to reuse existing geometry. Consideration must also be given to the modeling of the dependent geometry. For example, if a screw boss moves, the depth of the hole should change with it. Is there a modeling technique that could be used to allow the depth of the hole to automatically update?

Component Fastening

Plastic material can have a number of advantages over stiffer materials in some applications. The balance to be thought about is one of flexibility and performance. Chapter 2 covered a number of reasons you may have chosen plastic for your design. Whatever the reason, plastics designers are often in need of adding stiffening features to their component designs to increase overall component strength.

The strength of plastic parts needs to be considered in regard to the requirements of their assembled state (i.e., their function). For example, a Frisbee is a stand-alone part, whereas a five-component assembly requires that you consider the assembly's function when assembled. Such considerations affect the choice of how a product is held together. The following sections discuss methods of fastening plastic components.

Ultrasonic Welding

Ultrasonic welding provides a method by which plastic components are welded (melted together) to form a bond between the mating parts. Ultrasonic techniques are also used to stake items such as a steel weights or stiffener plates to plastic components. This technique works much like heat staking technologies.

The process of ultrasonic welding components together requires that one component be held firmly in a nest (fixture) and the mating component vibrated at an ultrasonic frequency by a horn (part of the equipment). Material is intentionally modeled in one of the components (energy director) such that it will melt once vibrated and form a bond between the components. During the vibration process, the pressure and an ultrasonic vibration is applied to one of the components. Once the cycle of melting with the energy director is complete, the assembly is held in the fixture until cool. This technique requires compatible materials for the bond to form. The horn is designed to closely conform the surface contour of the component being vibrated to ensure that the vibration is applied to the joint to be welded.

The illustrations that follow show the horn design. The first illustration shows the two components to be welded. The base would be held in a nest and the top housing would have the vibration applied by the horn (second illustration). An energy director (discussed later in this section) would be modeled on one of the components to provide the material melted to form the bond between the two parts.

Component Fastening

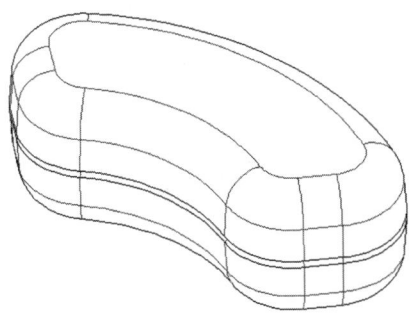

Components to be welded.

Horn designed to conform to the top housing surface contour.

➥ **NOTE:** *See Chapter 10 for other examples of fasteners.*

When components are ultrasonically welded, other components in the assembly may be affected. Are there internal components that cannot handle the vibration? Are there hold-down mechanisms inside the component that could also melt and relax the force upon whatever they are holding? Are there moving parts that could be melted and rendered inoperable? Most important, will you ever want to get the assembly apart again?

Ultrasonic welding also may not fit within a company's policy on recycling of materials or repair of product in the field. Your company may not want you to design something that cannot be taken apart for service or repair.

Consider, for example, the housing design shown in the following illustration and how Pro/ENGINEER might help you model the features required to seal the external portion of the product. The illustration shows an interlocking lip and support ribs added to help hold the interlock together.

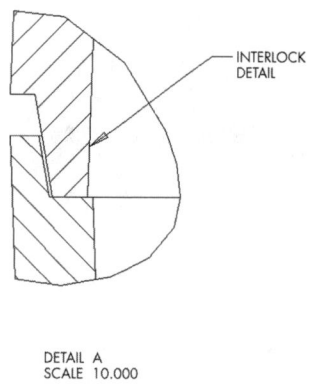

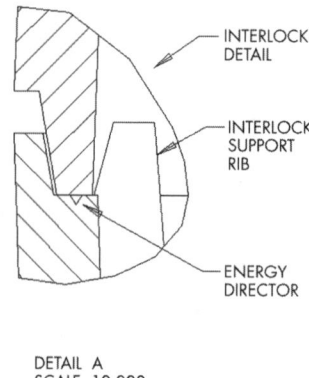

Interlocking lip around housings.

Modeled energy director, with support ribs added to help hold the interlock together.

As shown in the previous illustrations, the housing design has allowed for an interlocking lip between the two components. An energy director needs to be modeled on the mating face of the two housings so that there is material to form the bond between the components. During the welding cycle, the base of the assembly will be firmly held. Therefore, the horn needs to fit onto the top housing.

The energy director may be added to the lip simply by creating an Advanced Variable Section Sweep protrusion following the edge of the lip. A series of support ribs may be added to help keep the two components aligned during the welding cycle. The support ribs will also aid in the overall strength of the welded assembly.

Snap Fit Joints

Snap fit joints marry easily with the flexible properties of plastic material. However, when using snap fit joints, you need to consider the function of the snap joint. Pro/ENGINEER can aid in creating the snap geometry such that the geometry is easily modifiable and reasonably easy to reposition. The key elements in making the geometry easy to work with are to create the snap geometry in sequential order and to align to a stable reference only.

Component Fastening

Snap fit joints typically have to be fine-tuned so that you can try out the snap function to assure yourself that you are getting what you expected. This process normally requires simply tuning of the angle of engagement and the position of the seat when the assembly is snapped together. It is a good idea to have a discussion with your mold maker and to model "steel safe" in these areas until the first parts are molded. The tooling may then be dimensionally tuned to a final position.

Testing Snap Fit

Testing the snap fit may involve more than snapping a couple of components together. There may be standards you must adhere to, such as drop test requirements, that must be worked out through testing and that may force you to change the design of the snap shape. This can cause difficulty for the toolmaker. This is one part of a mold for which a mold insert might be in order. If the snap changes drastically, you can change the insert in the mold. Snap fit features are sometimes used in assemblies to position components and to hold them during the assembly process only. Once the main assembly is together, other features allow for proper component positioning.

Steel safe modeling requires that to modify the mold you have to remove material from the mold. You might choose to model the first release of your component to the toolmaker in an undersized fashion, or you can model the component full scale and communicate to the toolmaker the areas that must be steel safe. Determining which technique is the best practice might bring a lot of controversy to the department meeting table.

The design of the snap fit in plastic usually requires that you take advantage of the flexibility of the material. When components are assembled or snapped together, the force exerted to perform that function pushes the snap further into the mating feature than it actually needs to go to fasten. When a snap is assembled, the components rest in a preloaded (non-rest) form. Therefore, there is always some load on the snap joint, which helps prevent assembled components from rattling. The following illustration shows a snap fit tab assembly.

Snap fit tab assembly.

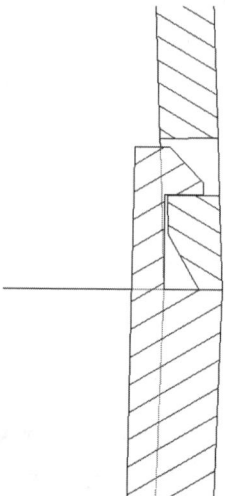

Modeling "Steel Safe"

Tooling is expensive. Part production tooling made from steel is usually hardened, making it difficult to modify finished tooling. Designing steel safe allows for a better situation if the mold has to be modified to bring a dimension into specification. Critical dimensions are usually highlighted as possible places where one would want the tooling for the part to be steel safe, thus possibly avoiding an expensive rework to the mold.

Steel safe is a term that applies more often to the tooling for a plastic part than to the plastic part itself. What does it mean to model something steel safe? Consider for example that you have a deep cavity in your plastic part design. When the mold for that component is created, there will be a protrusion on the mold to make the cavity in your part. If you were concerned about the depth of the cavity in your part and wanted to be able to fine tune the volume, you would have the molder make the cavity a little deeper than your design calls for so that you may be able to tune the depth after you see the first plastic parts.

In order to reduce the depth of the cavity after the tooling has been manufactured, the toolmaker would remove (grind or burn) material from the protrusion that made the cavity. The tool in the previous example was steel safe

Component Fastening 107

because the toolmaker actually had steel to remove from the protrusion in the tooling. However, what would have happened if there had been no material to remove (not steel safe)?

In the previous discussion, to make the cavity deeper, the toolmaker would have to add material to the protrusion in the mold (welding) or cut out that area of the tool (create an insert) and replace the protrusion. Both of these options are much more expensive than if the tool is designed steel safe.

Consider another example. In the following illustration, two piece parts from a product assembly have been put together. There is a housing (or holder) and a lever assembled to it. There would normally also be another part of the assembly used to clamp the pivot area of the lever assembled from the top. However, for purposes of clarity, only the housing and pivot lever are shown.

Housing and pivot lever.

The lever shown in the illustration is used to activate a switch at the other end of the part. There are three places where the pivot lever must be guided by the housings without restricting the ability of the lever to operate. The designer is concerned that the height positions of the three guides will need to be tuned such that they work harmoniously and do not restrict the movement of the lever. Another concern is the potential for the lever to rattle because it is too loose and cause the consumer to feel that the design is inferior. The designer has identified the three heights that need to be modeled steel safe. These are shown in the following illustration.

Critical height protrusions.

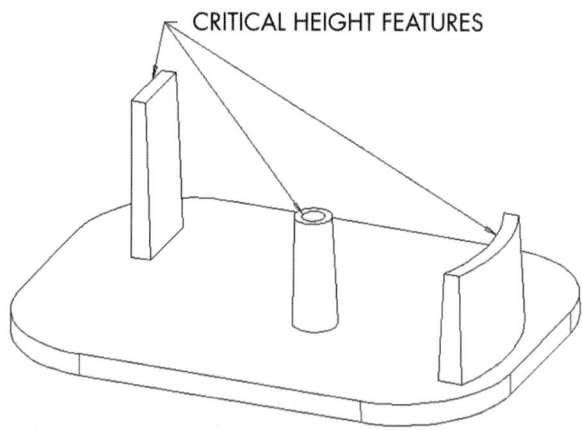

The three protrusions would be manufactured undersize to allow the tooling to be steel safe. After the first production parts are tested, the designer may wish to fine-tune the height of one or more of the protrusions. To do this would require the toolmaker to remove steel from the tooling (the tooling has cavities that represent the protrusions on the part), shown in the following illustration.

Tooling.

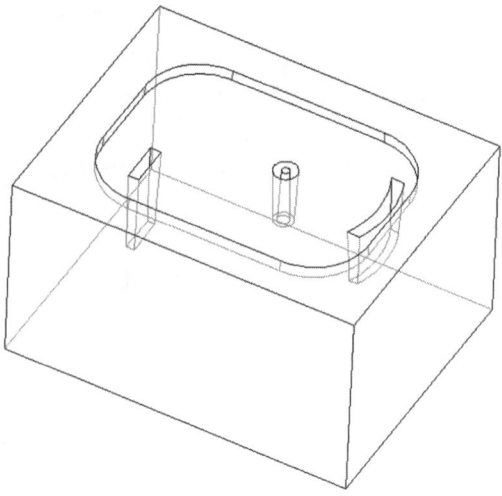

Component Fastening

A typical design of a component has the component modeled in its finished size. If areas of the design are identified as areas where you wish to stay steel safe in the tooling, there are two ways of handling this:

- Change the dimensions in the design file to be steel safe and issue the file to the toolmaker. If you do this, your design file does not reflect the design intent for your design.
- Convey to the toolmaker areas you wish to stay steel safe in the tooling so they may allow for it when they produce the mold.

Either way, you get a steel safe part. Company policy may force designers to model parts in their finished state only, so that design intent is not lost.

Using Screw Bosses

When designing screw bosses, the designer needs to consider the assembly process and determine what the finished assembly needs to offer the end user. Plastic components are designed to function in their assembled (restrained) position. As a result, and because of the flexibility of the material, if you measure two flat surfaces on parts you intend to put together, you will usually find that neither are flat unless pulled together by the assembly process.

This advantage (sometimes, disadvantage) of plastic does not normally cause problems for a design as long as the part meets the specification in the restrained position. Plastic parts should be measured in their working position to verify the dimensional stability of the model.

Consider the following example of a set of housings that clamp together using two screw bosses. The design intent is to hold down the electric motor using ribs internally on the housings during the assembly process to avoid having to add a separate motor mount bracket on the inside of the assembly. To take advantage of the flexibility of the plastics, the designer needs to consider what will happen when the screws are inserted into the housings at assembly time. The following cross-sectional diagram shows the assembly in final form. The motor is clamped inside the housings, the housings are mated, and the screws are properly seated in their respective recesses.

Sectional assembly view.

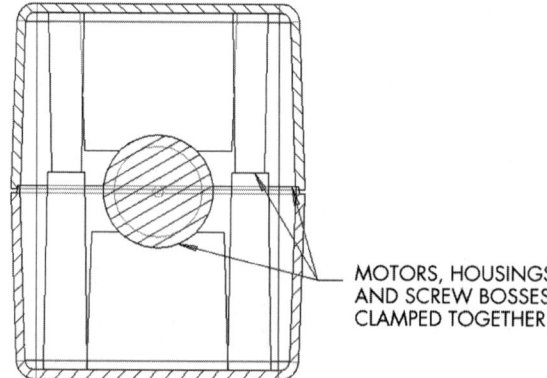

MOTORS, HOUSINGS AND SCREW BOSSES CLAMPED TOGETHER

If the components are molded as shown in the previous illustration, there is a high risk—because of manufacturing tolerance, part warpage, and so on—that the screw bosses will be seated before the assembly is pulled together by the screws. If, however, the design incorporates the traditional variances in the parts, the design can also help ensure that design intent is followed. If you want to ensure a satisfactory outcome, make a list of assembly requirements and model accordingly. Ask yourself what should happen during the assembly process, and in what order. In this example, the following should hold true as the parts are drawn together by the screws.

- The external profile of the housings should mate to seal the joint between the components all around the housings.
- The ribs that hold the motor in place should clamp onto the motor.
- The screw boss mating surfaces should touch.

In the following illustration, the geometry has been exaggerated to reflect the subject of discussion. It is noteworthy that the external interlock lip surfaces will mate first, followed by the clamp ribs, and finally by the boss faces.

Component Fastening

Exaggerated assembly sectional view.

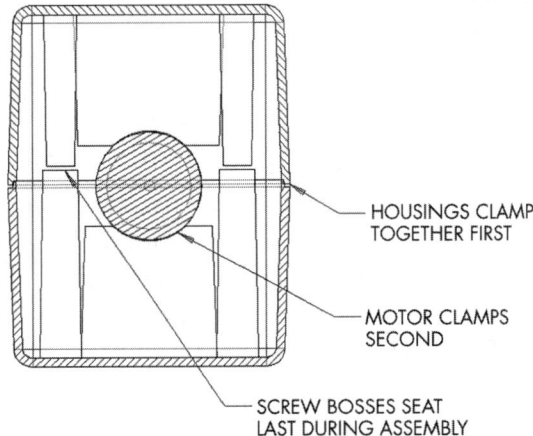

— HOUSINGS CLAMP TOGETHER FIRST

— MOTOR CLAMPS SECOND

— SCREW BOSSES SEAT LAST DURING ASSEMBLY

By following the clamping procedure previously described, the assembly will always be rigidly clamped together and you can therefore reasonably conclude that even if the parts are slightly warped they will conform to the design intent position. If clamping faces are modeled in Pro/ENGINEER as steel safe features, they can be quickly updated to fine-tune the design.

The screw boss design will accommodate some dimensional variation in finished components. Family molds in which two or four of the same components are made simultaneously will suffer from slight variations in finished component dimensions due to the variance in the cavity dimensions and the actual process of molding the components. Clamping components in this fashion will also overcome some of the instability between and among like components molded from different cavities.

The amount of gap left to be completely closed as screw boss mating faces are clamped together by assembly varies according to the material of choice and the overall size of the components. A screw boss gap can vary from 0.15 mm to as much as 1 mm (for large parts). Keep in mind that too much gap can result in easily stripped screws, major distortions of housings, and marking of the plastics due to stresses exerted on a product or part.

Wall Mount Holding Tabs

Wall mount holding tabs used in a design concept are also best handled in Pro/ENGINEER using offset datum planes from which to model components. The technique is much the same as the previous housing example. In this case, for example, a plastic part that holds something is intended for mounting on a wall. The chance that both the wall and the plastic part are flat and will mate perfectly with each other is rather remote. If the mounting tabs are offset from the perimeter of the component, the perimeter will be pulled tight to the wall during the mounting process, as indicated in the following illustration. The process will result in a much more rigid and perceivably higher-quality assembly of the unit on the wall.

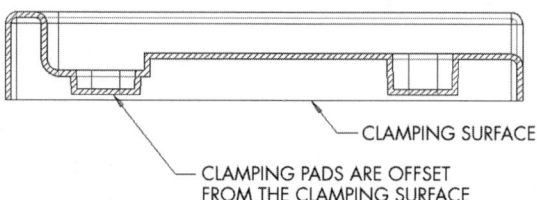

Wall mount bracket modeled to pull tight against the wall.

CLAMPING SURFACE

CLAMPING PADS ARE OFFSET FROM THE CLAMPING SURFACE

Summary

In this chapter you have learned techniques for identifying root features in the component design process. Logical feature creation was discussed to aid in organizing the modeling plan. How to keep track of buried features was discussed to make future changes to dimensions easier to deal with. Modeling plans were covered to help to develop the road map to follow when organizing and modeling features. The discussion on fastening plastic parts pointed out the overall design philosophy of component design. Part II continues the discussion in terms of individual features.

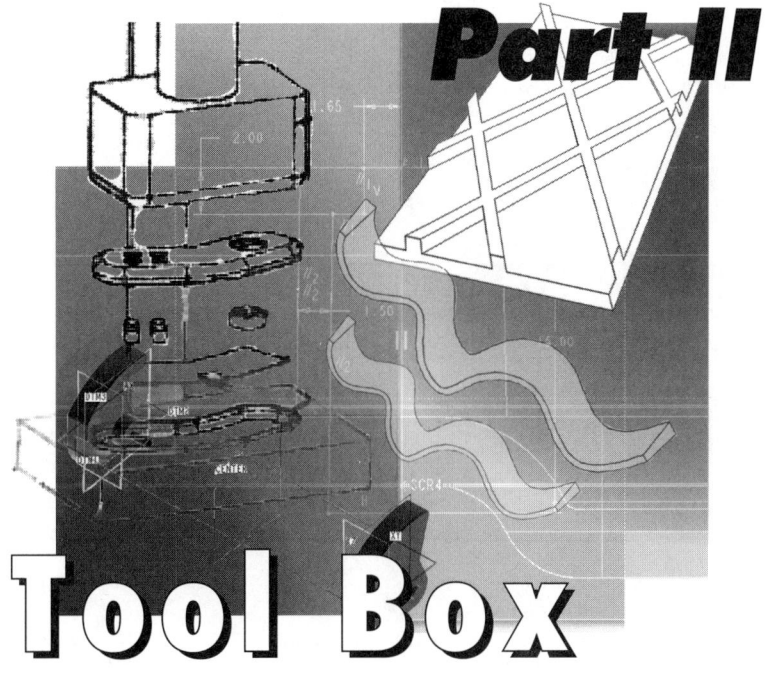

Part II
The Tool Box

In Part I, you were introduced to the basic concepts and fundamentals of plastics and plastics manufacturing methods, and to design philosophy and planning with Pro/ENGINEER as a design tool. Part II guides you through the use of Pro/ENGINEER features especially important to plastics design.

Part II deals with features and techniques commonly used in plastics design. The shell, which is a major Pro/ENGINEER feature and almost always used in plastics design, is discussed in terms of various applications of it. The many types of draft, another Pro/ENGINEER feature, are also major features employed in modeling. Attention is given to manufacturing issues, such as flash and weld lines, which can cause a plastic part to appear ugly or even make the plastic part weak. The three chapters of Part II describe how to use Pro/ENGINEER to create robust geometry by creating features that can be made and that are manufacturable.

Chapter 5

Plastics Design Features

Shells and Features That Aid Plastics Design

Introduction

Pro/ENGINEER provides features that allow designers to easily obtain very complex geometry. The Shell command, a Pro/ENGINEER feature, for example, will create an internal section of constant wall thickness based on external geometry. This is one of the most powerful features used in plastics part design. This chapter covers shell features, and cosmetic features such as draft and tweak offset.

Shells of Constant Wall Thickness

Pro/ENGINEER's Shell command enables you to create components of constant wall thickness, which are essential to plastic component design. Plastic parts require walls of constant thickness so that the parts do not shrink more in one place than in another. Unintended shrinkage causes sink marks or deformation areas in a part, which mar the final product.

Before the shell feature can be used, the designer creates the outside shape of a part that will eventually have walls. Just the basic outside shapes and contours of the product are created initially. Cosmetic features such as labels and small cuts are added later in the design process. These features are small and will not affect the shrinkage or any physical aspect of the plastic part. An example of an unshelled part is shown in the following illustration.

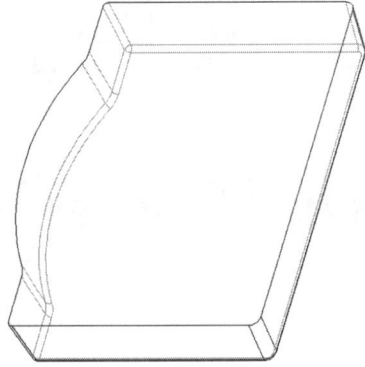

A part before shelling.

A part must be solid for the shell feature to function. The shell function will only work with solid features, parts, or models because the shell function must know which side of the surface is on the inside or the outside. Therefore, surface models that are not completely "watertight" (sealed at all edges of surfaces) or wireframe models cannot be shelled. If the surface model is completely watertight, Pro/ENGINEER can make it into a solid, and then the part can be shelled. The designer must fill in any spaces between extraneous surfaces and the part to make a solid model.

More than one surface of a model can be removed during the shelling operation. The designer determines which surface or surfaces of the solid model need to be removed. The designer must also determine the thickness of the part. Once this information is input, the part can be shelled. The following illustration shows an example of a shelled part exhibiting constant wall thickness.

Shells of Irregular Wall Thickness

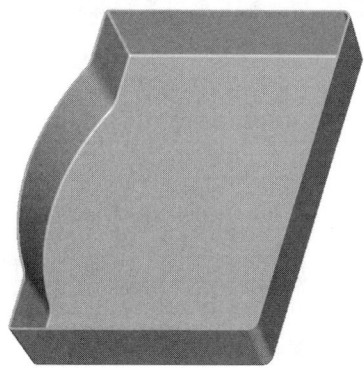

Shelled part of constant wall thickness.

Shells of Irregular Wall Thickness

Sometimes components require an overall constant wall thickness combined with areas or surfaces of a different thickness. Such surfaces, for example, might need to be thicker for the purpose of structural support, or thinner to allow for coupling with another component.

In the case of this type of component, the process for creating it is similar to that used to create a part of constant wall thickness. The designer starts with a solid model, then selects the surfaces to be removed, as was done for the constant wall thickness model. The difference now is that the designer has the option of changing the thickness of each wall of the part individually. Once the surface selections have been made, the shell can be created. The following illustration shows a shelled part exhibiting walls of irregular thickness.

Shelled part with walls of irregular thickness.

Manually Creating a Shell

When shells fail to build in Pro/ENGINEER, the designer must either identify and correct the problem area, or create the shell manually. Shells most often fail to build when the outer surfaces of a part are offset to the inner surfaces and the inner surfaces fold over each other as Pro/ENGINEER tries to create offset inner surfaces of the shell. In this case, Pro/ENGINEER does not know how to calculate parameters for the inside surfaces, and returns an error message stating that the part cannot be shelled. The following illustration shows a part for which shelling would not work.

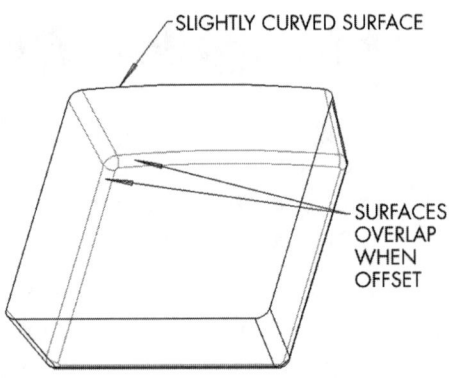

A failed shell section.

Typically, you do not know where the shelling has failed.

✓ **TIP**: *A quick technique of finding the area of failure is to make the shell thickness quite a bit smaller than your desired thickness.*

If, after you have done this, the component does not shell, look at the inside surfaces of the shell to see if there are any surfaces that have gone to a zero size. The following illustration shows the attempted shell of the failed part of the previous illustration and the surfaces that will fail because they are too thin.

Manually Creating a Shell

Part before the shell fails.

Once the bad surfaces are identified, the good surfaces must be selected and offset by the desired thickness. Each pair of adjacent surface edges must be extended beyond each other. The surfaces are trimmed back to the intersection of the surfaces. The extension beyond each other ensures that a good intersection will occur. Once all surfaces have been intersected, any open spaces must be filled with new surfaces.

The new surfaces can now be joined (as a quilt). The quilt is used as a surface cut into the solid Pro/ENGINEER part. The shell can then be created, as shown in the following illustration. Note that the inside of the part does not yet have rounds on its edges. The Pro/ENGINEER designer can now add the rounds at the required radii.

Completed manually shelled part.

Draft

Wherever possible, you should make draft for all features, such as bosses and ribs, in a part. Draft is the angle from a parting plane used as the plane for the perpendicular direction of the parting of the cavity and core in a molding machine. The following illustration shows a part with a parting plane and its draft angle.

A part showing the draft angle.

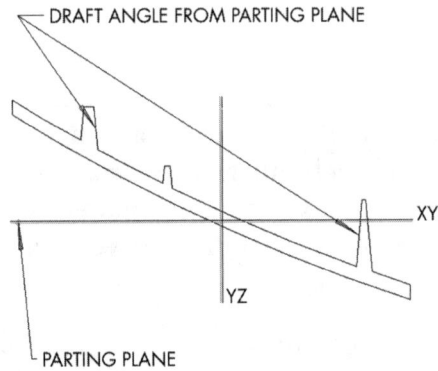

Draft is used to make a part manufacturable, and is critical in plastic component design because it allows a part to be separated from manufacturing tooling and molds. As stated in Chapter 1, draft is required for a part to be removed from a mold. Without draft, the part would stick to the mold and the stress of trying to pull the part from the mold would cause the part to stretch and break.

Draft in Pro/ENGINEER has some restrictions. Draft can only be formed from planes or cylindrical shapes. The direction of the draft angle must be made from a plane perpendicular to a reference plane. The reference plane in plastic design is most likely the parting plane.

Draft is used as a design tool in Pro/ENGINEER when, for example, the features of a cover and base need to fit together. The term *taper* is often used to describe a design feature such as this. For instance, a screw boss in the base of a part needs another feature on the cover of a part to help the part align itself when being assembled. Both parts must be tapered to allow for alignment at the first contact and a snug fit when the assembly is closed. The following illustration is an example of the use of a taper.

Tapered parts.

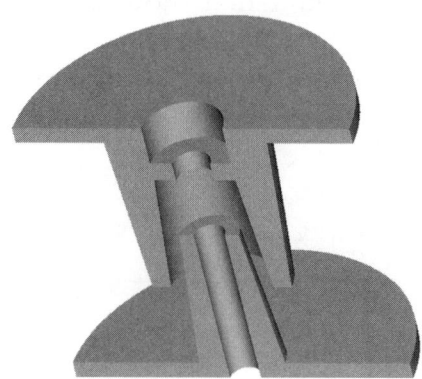

Pro/ENGINEER provides several methods of creating drafts. Draft angles in Pro/ENGINEER can range between −30 and +30 degrees. The plastics designer normally uses just the basic forms of the drafting feature. The draft features are described in the sections that follow.

No Split

A split is a location at which the draft angle changes direction. The location at which the draft angle changes direction might be a plane, a curve, or a surface. No Split is the most commonly used of Pro/ENGINEER's draft features. The designer wants features to draft in only one direction. Otherwise, a part will have undercuts that prevent the part from being removed from a mold. The previous illustration shows an example of a feature with simple (No Split) draft.

Split at a Curve or Quilt

The use of draft for split evident at a curve addresses primarily the needs of the manufacturer. With this type of split, draft is placed between the top and bottom surfaces of a part. Normal draft would start at the top or bottom surfaces of a part. The core and cavity must contact each other to ensure a good closure for the mold. Often the contact location needs to be hidden from view. Also, the mold maker wants to make the protrusions between core and cavity the same length. By doing so, the steel of which the mold is made will not be too long and fragile. Long, thin pieces of metal in molds are more easily broken.

The following illustration shows an example of using a curve and a parting quilt to create split draft. The reference plane for the draft is selected. Then the parting curve or parting quilt is selected. The parting quilt must go completely through the feature having the draft applied or the draft function will not work.

Split drafts.

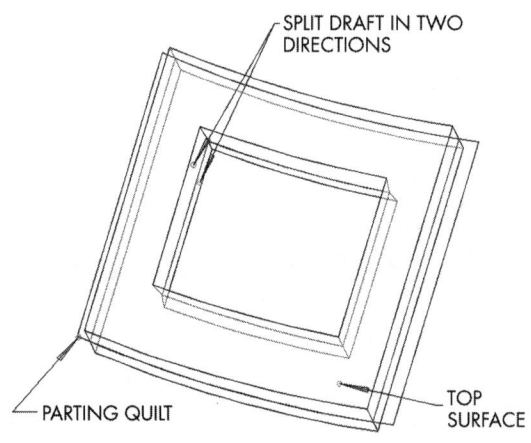

Using Split as a Sketch

Sometimes the space in a part is limited, which can cause Pro/ENGINEER design features to overlap. In many of these cases, the designer needs to be aware that the simple No Split draft feature cannot be used. The simple draft can cause interference with another feature, or a feature from the other half of the assembly may interfere with assembly.

By using the Split Sketch feature, a part can be made without causing interference, at the same time preserving the functionality required by the design. An example of a split sketch draft is shown in the following illustration.

Draft with the Sketch option.

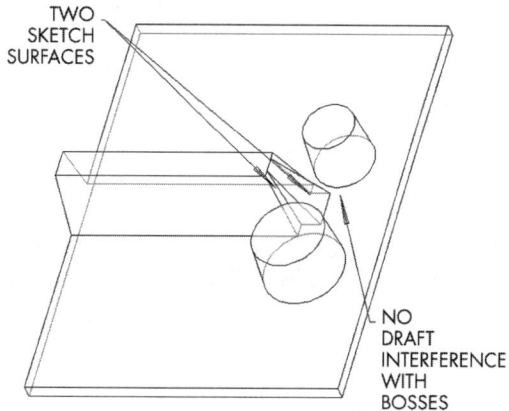

The draft feature can also be added to aid in the flow of plastic in a mold. The easier the flow of plastic in a mold, the stronger the part will be because of the consistent distribution of the material properties of the plastic. In the near future, Pro/ENGINEER will have protrusions that will build in the draft feature. This option will make the designer's job much easier because it will eliminate the need to add basic draft.

Features Aiding Plastics Design

A good plastics design provides features that improve the component strength of the final product and enhance material flow in manufacturing the product. The Pro/ENGINEER features described in the following sections aid in making plastics designs more robust.

Rounds

Plastic does not flow well around sharp corners. Therefore, a part design that requires sharp corners may call for a mold containing sharp corners that may not fill properly. Rounded (even slightly) corners allow plastic to flow and fill these areas. In addition, sharp corners are locations for stress concentrations. The stress concentrations can cause the part to fail under a heavy load or sudden shock. Round corners immensely reduce stress concentrations.

Another consideration in designing corners is that cosmetically the appearance of a part often looks better with rounded edges than with sharp edges. The use of rounds will often make a product more aesthetically pleasing. The sections that follow discuss the options available for employing simple rounds.

Simple Rounds

You should use simple rounds whenever possible. The computer calculation time is faster, and therefore the regeneration of features takes less time than for more complex rounds. Simple rounds also offer the advantage of a consistent radius. The user may select a single edge or tangent edges to make the simple round feature in Pro/ENGINEER. Four basic Pro/ENGINEER methods create simple rounds. These are discussed in the following sections.

Edge Chain Round

Edge chain rounds are the most simple to create. For this type of round, an edge is selected and the round radius is input. The round appears on the model, as shown in the following illustration.

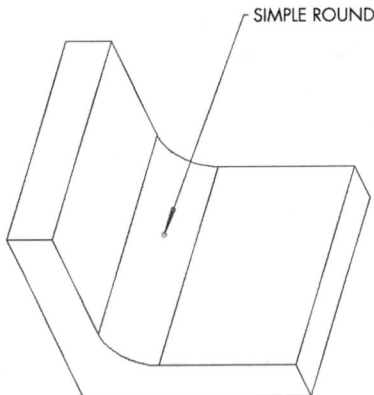

A simple chain edge round.

More than one edge may be selected at a time. The Pro/ENGINEER user must realize that all edges selected at the same time will have the same radius. If one of the radii needs to be changed later in the design, all radii will change.

Features Aiding Plastics Design

The designer must plan ahead to be sure that all edges selected in one sequence will have the same radii. Problems may arise if the model is passed on to another designer who does not necessarily realize that the rounds are related to one another.

Edge Tangent Chain Round

If the edges are tangent (that is, are at exactly 180 degrees to each other), the Tangent Chain option may be used. With this command, Pro/ENGINEER automatically goes from one edge to the next following the tangency rule. The user may find at this point in the design of the part that some edges are not tangent to each other. In this case, other methods of creating rounds will have to be employed.

For the model shown in the previous illustration, an edge tangent chain round can be made by selecting one of the edges of the model where the edges are tangent to each other. The following illustration shows the result of this selection, wherein all of the rounds have the same radius.

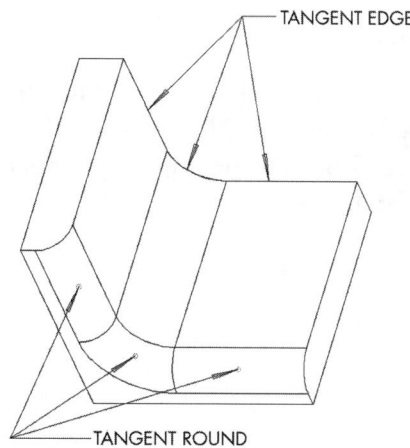

Edge tangent chain round.

If rounds are required around a surface, a surface may be selected. Simple rounds are created at the edges of the surface.

Edge Pair Round

When a flat surface needs to be replaced by a rounded surface, the Edge Pair round feature is used. With use of this feature, two edges are selected, as shown in the following illustration. This type of round is sometimes called a full round.

Edge pair round.

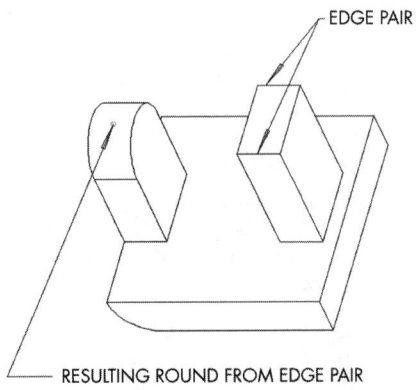

Edge-Surf Round

The Pro/ENGINEER user may want a round to go from one edge to an adjacent surface. In this case, the Edge-Surf feature is used. With this method, one edge is picked and then the adjacent surface is picked. These features are shown in the following illustration.

Edge-to-surface round.

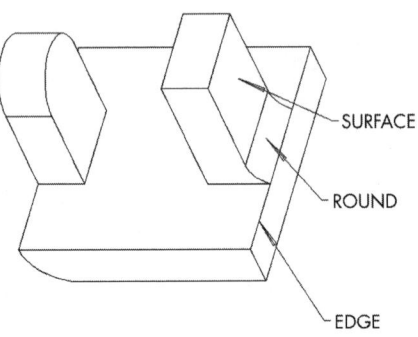

Surf-Surf Round

Sometimes it is easier to pick adjacent surfaces to create rounds. In this case, Pro/ENGINEER's Surf-Surf feature can be used. By selecting one surface and then the adjacent surface, Pro/ENGINEER creates a round along the edge between the surfaces. The following illustration shows an example of the outcome of this method.

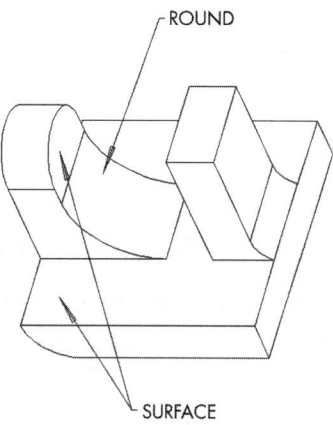

Surface-to-surface round.

Automatic Blending of Rounds

A part might have two surfaces, one of which is shorter than the other. For example, a small Pro/ENGINEER feature might be found on a larger feature, such as a protrusion extruded from a plane. This is shown in the illustration that follows. In this situation, the designer needs to decide how to terminate the rounds at the edge of the smaller feature's shorter surface.

Pro/ENGINEER offers the user two options for dealing with this situation. The round may be completely stopped at the end of the shorter surface using the Term Surfs option. Alternatively, the user can employ the Auto Blend option to create blend transitions where non-tangent edges exist. The following illustration shows examples of automatic blending of rounds. The surface-to-surface round must be selected for this feature to work.

Auto-blended rounds.

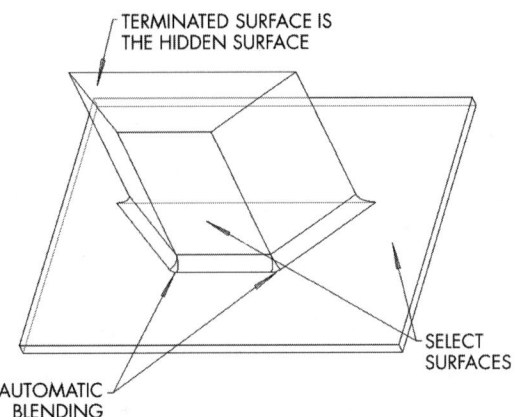

Simple Variable Radius Rounds

In Pro/ENGINEER, simple rounds can have variable radii. This makes it possible for the designer to allow the radius to change constantly along an edge.

For aesthetic reasons, the designer might vary the radius along an edge by adding additional points, where different radius values can be placed by Pro/ENGINEER on the edge. The designer might want to do this in a situation such as having light reflect back from the edge at a different roundness. This would give the user of the product an idea of where an important function of the product may be located, without labeling the area in question. The following illustration shows an example of rounds of varying radii.

Rounds of varying radii.

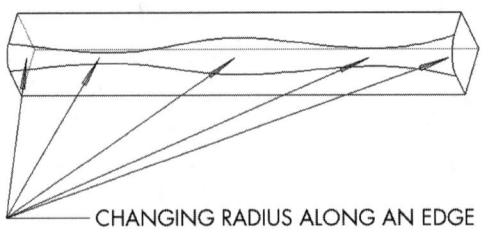

Advanced Rounds

Quite often in plastics design, features are located so close to each other that simple rounds will not work. If a simple round crosses into another feature, or if the material is too thin for the radius of the round, the round cannot be made. In these cases, the designer can use the Advanced round feature in Pro/ENGINEER.

An advanced round has all the functionality previously stated for simple rounds. In addition, the designer has more options available with the Advanced round feature to make changes during the creation of rounds. Advanced rounds in Pro/ENGINEER give the user the choice of adding sets of radii when making a round. The sets contain information about the type of round; for example, whether the round is to be of constant or variable radius.

An advanced round can contain many sets, where each set defines the shape of an individual round. The sets may be removed, added, or redefined at any time. The definition and changing of these sets is found in the Advanced round feature of Pro/ENGINEER.

Many options are available in Pro/ENGINEER for creating the desired shape of an advanced round. The designer can make spherical, swept, and patched rounded corners. Each of these corners may have radii of the same or different values. The following illustration shows an example of the corner types obtained when using sweeps and spheres.

Various corner shapes generated using advanced rounds.

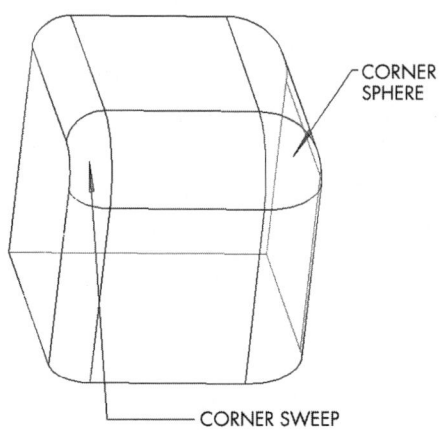

Advanced rounds are used for features for which simple rounds would fail. The following illustration shows two methods of making a round pass by a cut in a part. Part A makes the round seem as if it were placed there before the cut in the part. Part B shows a round that looks like a ball rolling along the corner edge.

Rounds passing by a cut feature.

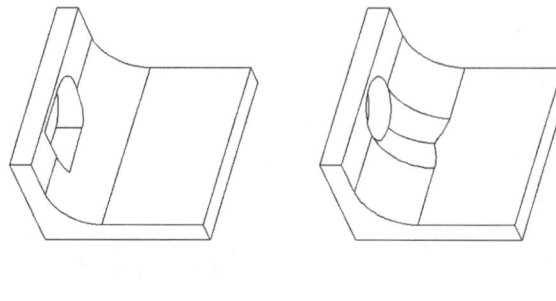

PART A   PART B

Crowning

Large, flat surfaces have a tendency to warp due to internal stresses developed when plastic was flowed through the mold that produced them. To mitigate against such warpage, the best option is to make the initial feature a curved protrusion. This helps by having the curved shape resist any internal stresses that may want to deform the part. The designer creates the protrusion feature of the part curve in the direction needed to meet the product's design intent.

Another option to help solve the warpage problem is Pro/ENGINEER's Radius Dome feature. A radius dome acts similarly to an eggshell. An eggshell is difficult to break and is very strong in compression. The shape of the dome makes it difficult to bend, warp, or push in the surface. The radius used under this option must be mathematically quite large compared to the dimensions of the surface's edges or the shape cannot be made. The dome need not be prominent. An example of a radius dome is shown in the following illustration.

> **NOTE:** *The radius dome should be added at the very start of a part, when the side edges are still straight and flat. The radius dome works on curved edges, like cylinders, but the edges are raised higher than the original cylinders and this may not be the effect you want to achieve.*

Radius dome.

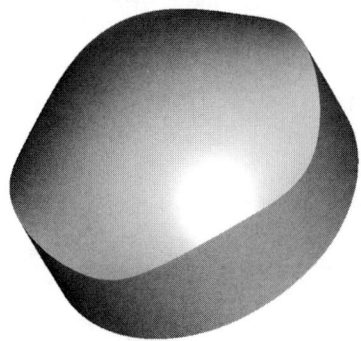

Local Thinning and Thickening of Material

Pro/ENGINEER provides users with tools for increasing or decreasing material thickness in a localized area. This is a feature especially useful to the plastics designer. For example, company names and labels are often added to parts using this feature. An example of the use of this feature is shown in the following illustration. In this example, the shapes of the letters ON would be transferred from a sketching plane onto the surface. The sketching plane would have to be parallel to the parting plane of the part in order for the part to be removed from the mold.

Tweak offset feature.

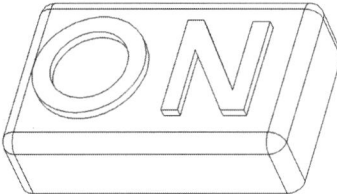

Ears

An ear is a protrusion that extends from the end of a surface and bends away from the surface at an angle. There are two types of ear features: variable and tab. The tab ear is always bent at a ninety-degree angle to the surface. The length of the tab is the only changeable dimension needed. The variable ear has both length and bending angle as changeable dimensions. The following illustration shows a Pro/ENGINEER ear feature. These two features can be used to lock or latch components of two parts together.

Variable ear feature.

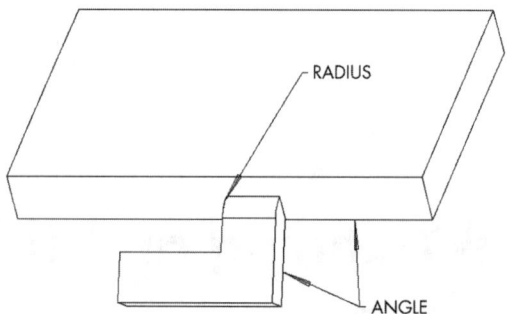

Lip Feature

The plastics designer needs a method of creating a feature that aids the contact and location of base to cover when products that contain such components are used. This feature, called a lip, and its use are covered in greater detail in the projects discussed later in this book. A lip is a protrusion that mates to a surface and is driven along the continuous edge of the surface. The following illustration shows the basic Pro/ENGINEER Lip feature.

Lip feature.

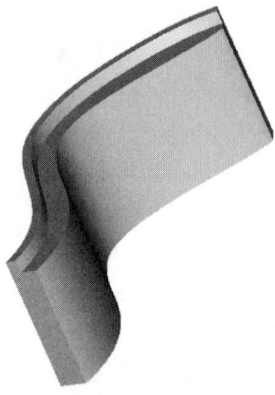

Replacing a Surface

A surface in a part may need to be replaced by another surface. Pro/ENGINEER allows for this through a feature called Replace, found under the TWEAK menu. The user begins by creating the new surface. The user then selects from the REPLACE menu the surface of the solid to be replaced. The new surface is then selected, which replaces the old surface. This feature is used when no other feature for making the contour of the surface is available.

Reminders About Mold Creation

The designer of plastic parts should keep in mind the following points about the mold creation process.

- The texture of a part, whether or not shown on the actual 3D part, must be considered in terms of draft.

- The parting line for the mold is not really a line that needs to be shown in Pro/ENGINEER. It could be a series of surfaces where the core and cavity meet in the mold.

- Inserts in the mold are not shown in Pro/ENGINEER. The designer must remember that undercuts in the design may require inserts and must prevent other features in the design from interfering with the insert.

    NOTE: *A mold maker may need to be consulted to determine the best insert design.*

- Runners and gates are not shown in Pro/ENGINEER. The designer must be aware of the possible locations of gates. The design must be made to help aid the flow of the plastic through the part. An example, a rib not going into an exterior wall, was discussed in the section on ribs.

    NOTE: *The designer might be well advised to consult the mold maker for the best gate locations before finalizing the design.*

Summary

Pro/ENGINEER includes many features that address the plastics designer's needs. The use of shells, draft, rounds, crowns, domes, lips, and special tweak features make the design of plastics much easier. You have seen in this chapter how the shell feature is used to make a solid part into a part with constant or varying thickness of walls. You were also shown a method of manually creating a shell, using various Pro/ENGINEER features, for situations in which Pro/ENGINEER cannot make the shell automatically.

Draft features were described, including their purpose and where they would be used in the design of a plastic part. Rounds were separated into simple and complex features, and discussed along with a sample of the round features a plastics designer might use. Domes and crowns where described as methods of making flat surfaces round and to eliminate the risk of parts warping due to internal plastic stresses created by the various molding processes.

The Tab and Ear features were described as features that may be used to create attachments of one part to another. The Lip feature, which is very important in plastics design, was introduced.

> **NOTE:** *The Lip feature is explored in more in-depth use in Part III.*

The next chapter explores employing the features discussed in this chapter toward providing strength in plastic parts.

Chapter 6

Strength of Shapes

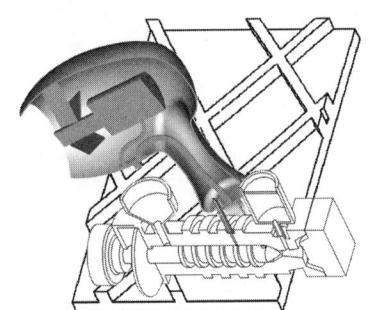

Introduction

The shape of plastic parts and components in products can make a product stronger or weaker. In this chapter, you will discover methods of changing the strength of a design. This chapter also discusses how and why features are created to strengthen a part.

Component Shape and Overall Part Strength

Stresses that affect plastic parts are heat, bending, deflection, and shock. The shape of a part can increase or decrease the effect of these stresses. This section describes very basic guidelines for designing in such a way that you minimize the negative consequences produced by such stresses.

Solving Heat Dissipation Problems

As was stated in earlier chapters, plastics lose their rigidity, and thus strength, as temperature increases. The following illustration indicates how increasing temperature affects the strength of plastic.

Effects of temperature on plastic strength.

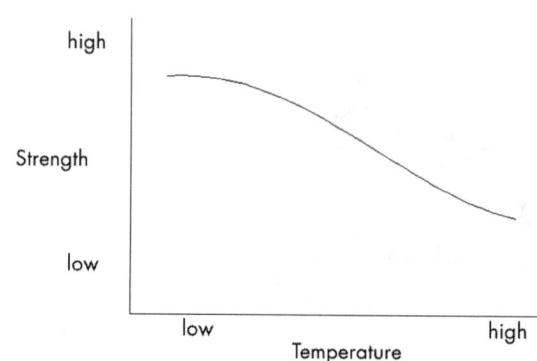

Plastics can be designed to aid in reducing heat transfer from one area to another. An example would be to have a plastic barrier shielding a user from the heat produced in an audio amplifier. This example is discussed later in the chapter. The thermal conductivity of plastics is lower than that of metals. Therefore, plastics act as a better shield against heat transfer than metal. The thicker the plastic wall, the more of a temperature gradient across it, and thus the more that it will shield against. However, a thicker design means that a part will be heavier and will cost more money to produce and market.

Air is a very poor conductor of heat. If the user recognizes this fact, Pro/ENGINEER can be used to make a better design for a plastic thermal wall. The following illustration shows an example of how to use the shape of a wall to have it act as a better thermal barrier.

Component Shape and Overall Part Strength

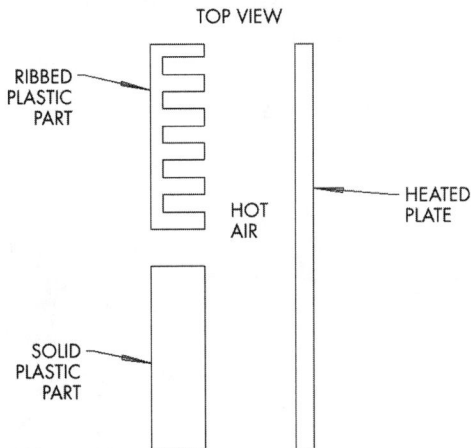

Two designs for thermal walls.

In this illustration, the part containing ribs is smaller in width than the solid model. However, because of the poorer thermal conductivity of air, less heat is transferred through the plastic wall. The rib design is better than the thicker solid model as a thermal barrier. Plastics are often used to surround objects that give off heat. For example, an audio amplifier might have a plastic outer shell, as shown in the following illustration.

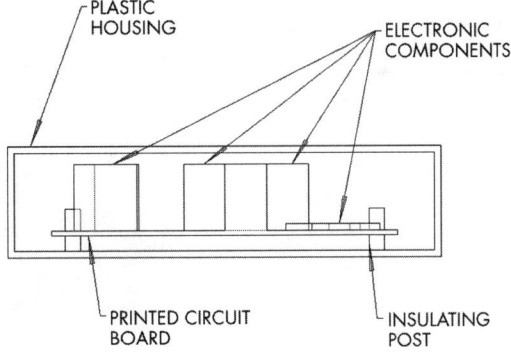

Cross section of an audio amplifier.

The plastics around the amplifier do not touch the hot electronic components. Insulating material holds the printed circuit boards, and the plastics hold the insulating materials. Therefore, the plastics do not melt. However, the air inside the amplifier gets very hot. The hot air can melt even high-temperature resistant plastics. The designer must incorporate features that allow air to escape the amplifier. The first thought is to incorporate vents at the top and bottom of the amplifier. The top of the amplifier is shown in the following illustration.

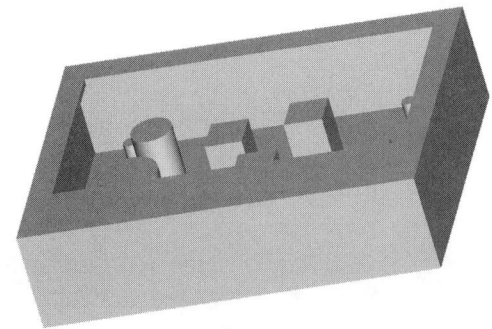

First try at cooling an amplifier.

This design lets the air out, but the wall is weak structurally. In addition, safety problems arise because putting a finger through the large gap may burn or electrocute the user. The next solution is to design louvers. A louver is an air gap between two solid parts. To use this concept in the amplifier design, many louvers are placed in a row, as shown in the following illustration.

An amplifier wall with louvers.

This design allows enough air through the amplifier to keep the plastic below its melting temperature. In addition, any safety hazards have been avoided by the addition of the louvers. The designer can create these louvers quite quickly in Pro/ENGINEER by using cuts and patterns.

Component Shape and Overall Part Strength

By solving the heating problem indirectly, with the flow of air cooling the hot components in the amplifier, the designer is able to incorporate a plastic part as the shell of the device. A direct method of cooling the hot components would, for example, be a metal heat sink connected to the hot components, with the heat sink visible and touchable on the back of the amplifier. Because the plastic is cooled by the air during operation, the consumer will not feel the full effect of the heat on the outside of an amplifier and will not get burned.

Guidelines for Bending and Deflection

Plastics can be designed to bend or to resist bending. To design a part that needs to bend, the designer should make the part with a low profile, as shown in the following illustration. This is done by incorporating a low cross section in the plane of the part that needs to bend when the product is used. The following illustration shows a low cross section, which allows for bending.

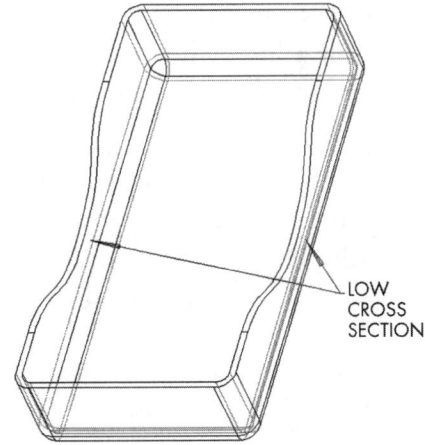

Low cross section for ease of bending.

Conversely, adding height along sections of the bending plane of a part increases resistance to bending. This is called a high profile, shown in the following illustration.

High profile in a part for resisting bending.

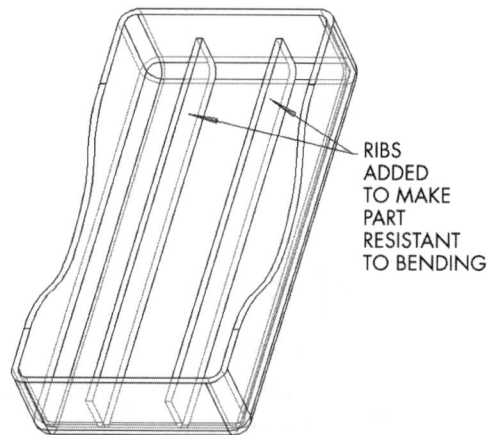

RIBS ADDED TO MAKE PART RESISTANT TO BENDING

Notice that the walls of the part are straight. The higher the walls, the more resistance there is to bending in the bending plane. This feature is easy to make in Pro/ENGINEER, but is not always what the product requires. The walls may have to be of a lower height for the product to fit into other parts, or a high profile might simply be cosmetically unacceptable.

In these situations, the designer must determine a method of achieving the same resistance to bending within the limited height. One solution, described in terms of use of the Radius Dome feature in Chapter 5, is to use rounds. Rounds spread the tension and compression stresses of the bending force over more of the part. Therefore, instead of stress concentrations at the corners of vertical walls, the stress is spread over a larger area, with more plastic to resist bending. The following illustration shows an example of two shapes that resist the same amount of bending force. The round edge is smaller in height than the sharp corner.

Component Shape and Overall Part Strength

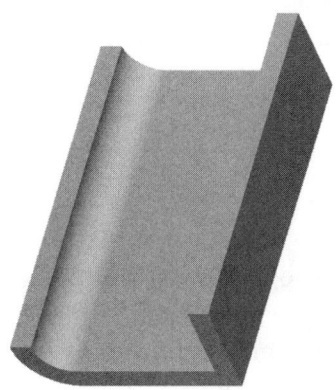

Two profiles to resist bending.

Quick Impact or Shock Guidelines

Shock problems are very difficult to solve completely. For example, when a handheld calculator drops, as shown in the following illustration, and hits the floor, the user does not want it to fly apart.

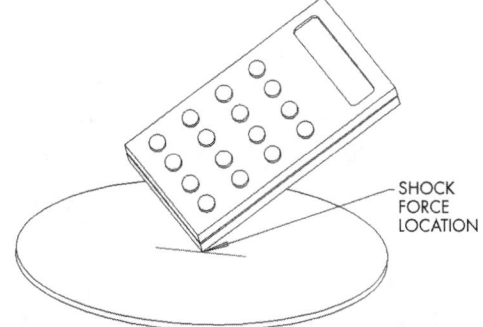

Calculator under shock forces.

The shock occurs at the moment and a few milliseconds after the part hits the floor. The kinetic energy is transferred into sound energy and potential energy (as bending of the part). The part needs to be designed so that it will withstand shock forces in all directions the designer thinks the part might be struck. That is, the calculator needs to withstand shock forces if dropped on any of its eight corners or any of its surfaces.

The weight and position of the materials within a part also affect the transfer of kinetic energy to the plastic components of a part. An example is the battery within the calculator. A part should be designed so that heavy parts within it cannot move relative to the housing. Protrusion features are used to secure interior components and minimize their movement.

Protrusion features are typically designed in such a way that heavy objects fit snugly within them. In this type of design, the plastic housing and the heavy weight then form a solid block capable of absorbing greater shock than if they were separate objects smashing into each other. The energy is transferred around the weight and fitting. Otherwise, the weight would put high forces on the plastic supporting structure during the shock and the weight would break the supporting structure.

A rigid design has many features that contact one another on the two housing sides. If a part is made too rigid, the energy from a drop is placed completely on the features that connect the two sides of the housing. Unless the features are strong enough, the part will break at these contact points.

You need to design in such a way that the energy from a drop is not completely transferred to connecting features. Instead, the housings of a part should absorb some of the energy by bending and recovering, releasing the otherwise destructive energy back to the environment. In this type of design, the housing can move and return to its original shape, and connecting features do not have to be as rigid or placed as closely together (more of them) to do their work.

Rounds on the corners of parts help reduce the stress of impact. For example, your computer mouse and your handheld calculator have rounds on their corners to reduce point contact (the surface area contacting the object impacted with), and therefore the stress concentration, during impact. The following illustration shows a better design for the calculator than shown previously.

An impact resistant calculator design.

Rib and Web Design Techniques and Rules

Ribs are used to add strength to a component. The location of ribs is an important consideration for the plastics designer. In Pro/ENGINEER, a rib is a feature that creates a small or thin extrusion from a surface of a part. Ribs are attached to a part's existing design geometry. Ribs in Pro/ENGINEER must be tied to existing geometry to function. If they are not, they will fail.

In addition, a Pro/ENGINEER rib cannot be added to empty space, where the rib's basic function of strengthening the part is lost. For example, when you use the Pro/ENGINEER sketcher to create a rib, the rib sketch must start and stop at the surface of the part. Pro/ENGINEER will not let you create a rib feature that starts at a surface and stops in empty space. Pro/ENGINEER incorporates two basic types of ribs: straight and rotational. These serve different purposes, which are explained in the sections that follow.

Straight Ribs

To create a rib in Pro/ENGINEER, you must first draw an open section on the sketcher. Remember that the ends of the open section must be terminated at a surface. The following illustration shows an example of a straight rib.

Straight rib.

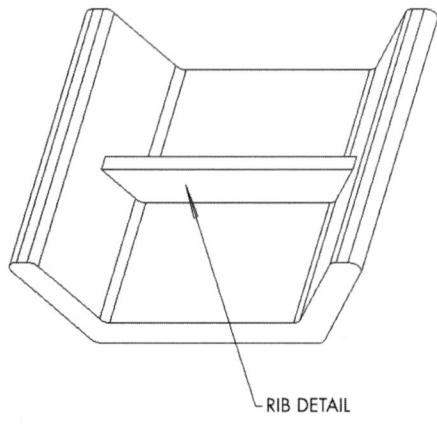

Rotational Ribs

Rotational ribs are added around a part's bosses and cylinders because they add a curved surface to the end of the rib farthest away from the boss. A straight rib would add a straight surface. To create this type of rib, you begin by creating a sketch of the side view of the rib on a Through/Axis datum plane. You then input into the sketcher the thickness value of the rib. Pro/ENGINEER then rotates the sketch about the axis. Two parallel planes, separated by the thickness of the rib and centered about the sketching plane, cut the rib. The result is a rib with parallel sides but a curved surface at the open end of the rib, as shown in the following illustration.

A rotational rib.

Techniques and Rules

In summary of the previous sections, there are several points you should keep in mind in regard to ribs. Ribs are used in plastics design to add structural strength to components and parts. Ribs decrease the flexibility of the surface of a part, and are usually placed inside a part because they tend to detract from the appearance of a product. The following design and technique rules should be followed when creating ribs.

- *Keep the profile as low as possible.* The lower the ribs are, the less chance of them causing interference with parts in the model. In addition, the plastic material will have a difficult time flowing during the molding of the part if the ribs are very tall and narrow.

- *Make the rib wall thickness 50 to 60 percent of the outer wall thickness.* Sink marks are likely to appear if you use a rib wall thickness greater than this value. The plastic between the base of the rib and the wall is thicker than the outer wall thickness. This plastic will take longer to cool and will shrink more than the rest of the wall.

- *Use a web design to resist twisting, and increase torsional stiffness.* The following illustration shows a web design using ribs. The design prevents the part from warping after being removed from the mold and prevents the part from being twisted.

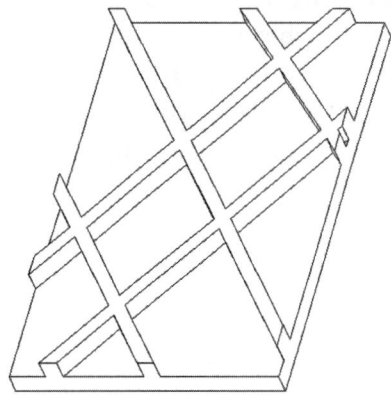

Ribs and outer walls.

- *Keep ribs away from exterior walls.* Ribs are helpful on flat sections and at the center of parts. A good rib design creates ribs that are brought up to but not quite touching an exterior wall, as shown in the following illustration. This rib will perform its original strengthening function and decrease turbulence during the flow of the melted plastics along the outer wall and rib. This reduces residual stress areas between the rib and the wall, therefore making the final product stronger and more resistant to impact and bending stresses.

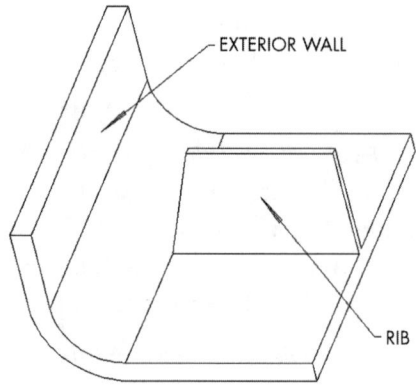

Ribs and outer walls.

Ribs and Draft

Once ribs are constructed, draft must be added to them. Ribs should be treated differently from other features in a part when it comes to adding draft. Keep in mind that ribs are usually for structural additions. Therefore, ribs do not have textured surfaces. Because they have smooth surfaces, ribs do not have to have a 2- to 5-degree draft angle. Often a 1-degree draft angle is sufficient. The following illustration shows what surfaces to pick when adding draft to simple ribs.

Adding draft to simple ribs.

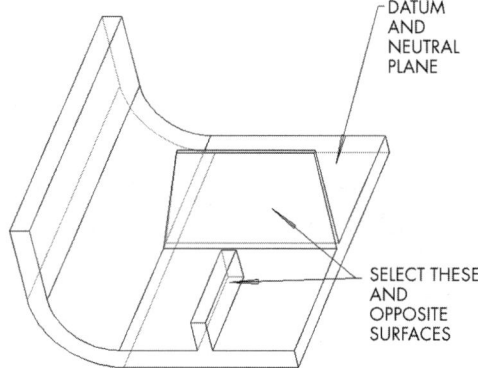

If a rib is tall, and a draft of 1 or 2 degrees causes the rib base to be greater than 60 percent of the thickness of the wall to which it is attached, you must find another solution. Two possible solutions are shown in the following illustration.

Two alternate rib draft solutions.

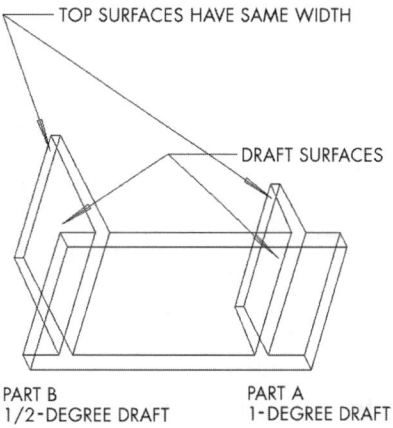

The successful part A solution in the previous illustration starts the rib at 60 percent of the external wall and grows the rib up from the wall. A 1-degree angle was needed. In the part B solution, the top part of the rib came to a point. The part B solution design is not acceptable. The designer may have to go to a $1/2$-degree angle or no draft angle at all to make this feature. At this point, the designer must consult with the mold maker to determine if the design is feasible for the part B solution.

Curved Ribs and Draft

Straight ribs are relatively easy to draft. Curved ribs are more difficult. To draft curved ribs, you need curve-driven draft that follows the rib contour, providing an even thickness at the top of the rib. The top of the rib is chosen as the curve from which the drafted surface will be generated. By choosing this curve, the rib will be consistently the same thickness at its top. A draft angle is input and Pro/ENGINEER creates the draft at the appropriate angle, as shown in the following illustration.

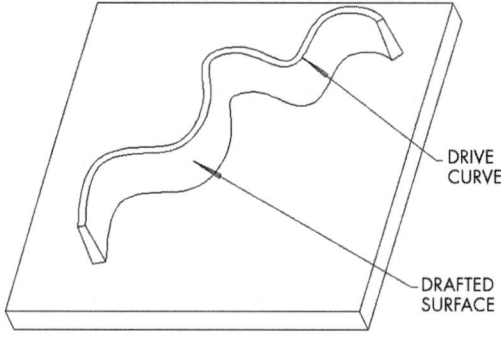

Draft on a curved rib.

Curve-driven draft can be used on parts that are not rectangular in shape and that have sculptured surfaces. Many times this type of rib is the best feature to use as a strengthener.

Ribs and Thin Wall Designs

Chapter 11 explores thin wall designs in detail. For the purposes of the current discussion, a thin wall design is one that incorporates walls that are thinner than the standard wall thicknesses elsewhere in the design. Ribs can be designed so that they have the same thickness as the walls on thin walled parts. The advantage of thin walls is that there is less plastic material to cause shrinkage in a part when it is removed from a molding machine.

Other Applications for Ribs

Ribs can be used to hold down other parts in a product. For example, a battery placed on one half of a container in a device can be held in place by a rib on the other half of the container. Ribs can also be used as interlocking devices. When two parts are being assembled, ribs can be used to guide the two halves together. Once the halves are assembled, the ribs hold the halves in place. This can help prevent rattling of components, which can make a product sound and feel cheap. The following illustration shows a hold-down feature and an interlocking rib.

Other uses of the rib.

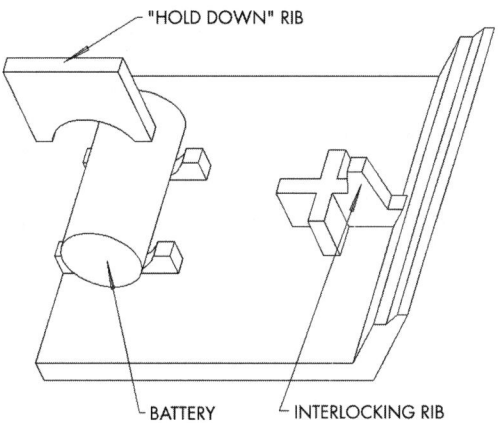

Summary

Pro/ENGINEER can be used to greatly aid the designer in making the design of plastic parts stronger and safer. The parts can be designed to allow for heat, shock, bending, and other stresses. Plastic parts can be used as thermal insulating walls and as safety barriers. Plastic parts can be made stronger through the following methods.

- By adding ribs
- By placing ribs at the appropriate locations to reduce twisting and torsional effects

- By modifying the height of walls
- By adding rounded corners to parts
- By using ribs to "hold down" parts
- By using curve-driven ribs that add strength while following parts that are not rectangular
- By using ribs as interlocks to keep parts together

The designer then ensures that these design features work within the context of a manufacturable product, which is the topic of the next chapter.

Chapter 7

Designing for Manufacturability

Introduction

Almost any shape can be created with a computer-aided design tool such as Pro/ENGINEER. A shape might look great, with its flowing curves and appealing sloping surfaces, but it might not be manufacturable. The are numerous reasons for this. This chapter is intended to help you create components and parts that are manufacturable, or more easily manufacturable, using Pro/ENGINEER.

Modeling Techniques That Enhance Material Flow

Material flow is a difficult problem. It has been a black magic area for years. The mold maker used to rely on experience gained from failures and successes in producing good products. Only in the last 15 years have the computer and material properties sciences been combined toward reducing the trial-and-error mystery of material flow. This section covers basic methods of designing parts for molding that introduce the material flow problem.

The designer must be conscious of the need to create parts for which the flow of plastic material in the molding process is maximized. The following guidelines describe what shapes are best for plastic design in terms of material flow, as well as problems that can arise related to component and part design.

- *Smooth walls reduce shear stress as plastic flows through a mold.* The reduction in shear stress means that the plastic will flow evenly along the entire edge of a part. The surface stresses that form during the molding will be constant completely around the part. Therefore, the part will not bend, twist, or warp after the molding process due to surface stresses. Rough walls, bosses, holes, and ribs in the flow path contribute turbulence to the flow, which can cause uneven stresses throughout the model that in turn cause warpage.

  > **NOTE:** Turbulent flow during the molding process causes areas of uneven stress in the final product after the part has been removed from the mold and the plastic has cooled down. The internal stresses then cause the part to warp. The warpage can happen immediately or gradually over months after the part comes out of the mold.

- *Design the part to allow the plastic to completely fill the mold under standard injection pressures.* As plastic flows through a mold, the front of the flow is parabolic in shape. The material fills the mold space completely and evenly as the plastic flows through it. Cooling and solidification of the plastic occurs evenly from the walls of the mold to the interior of the material. Because the plastic cools and solidifies from the walls to the center, eventually the cooled, solidified plastic beginning at the walls and working inward will meet at the center and stop the flow of hot plastic past this point.

- *There is limit to the distance plastic can flow through a mold before the solidification of the plastic between two walls stops the flow.* The term used for this distance is *flow length*. Once the flow length is reached, the plastic can no longer flow and fill up the remaining part of the mold. In a good design, this

Modeling Techniques That Enhance Material Flow

limit is never reached. Keep in mind that the thinner the walls, the shorter the flow length. Also, the more viscous the plastic, the slower it will flow, cooling down more rapidly and shortening the flow length.

Injecting plastic at a higher pressure, or increasing its temperature, can increase flow length. Both of these methods allow viscous material to reach farther into a mold.

> **NOTE:** *Higher-tonnage mold machines are required (to hold the mold together) as injection pressures increase.*

- *Design the mold to allow trapped air to be released.* For example, as plastic flows along a tubular shape in a mold, all of the air is pushed out of the tube. The air cannot get trapped in any nooks or crannies and cause a short shot. If it is not possible to push all of the air out, vents need to be designed as part of the mold. However, do not consider trapped air something that should halt the design process. A problem with trapped air can be solved later in the design process.

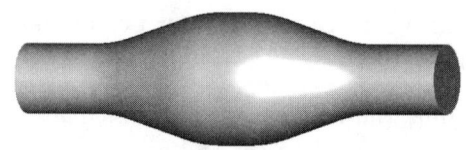

A shape that enhances material flow.

Viscous plastic flows most easily along a rod with entrance and exit of conical shape, as shown in the illustration at left. This is the ideal shape for material flow, but obviously all products cannot be shaped like this. In contrast, the following illustration shows a design likely to produce trapped air, as well as create turbulence in the material flow. The air gets caught in dead-end sections where the part is last to fill.

A shape that creates turbulent material flow.

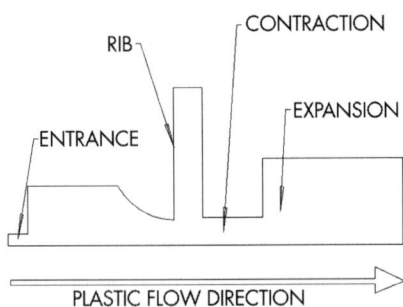

The shape shown in the previous illustration causes turbulent material flow for the following reasons.

- *The shape causes turbulent flow, which in turn causes warpage of the part.* A sudden expansion occurs at the entrance of the shape. This causes turbulent flow because the entrance begins small and rapidly expands, which creates a situation in which surface stresses vary around the entrance. Turbulence also occurs at the rib section and the expansion section. All of these effects cause variable stresses on the surface of the part, which in turn can make the part warp and bend when produced.

- *The shape will not fill under constant pressure, which will cause uneven internal and surface stresses along the part.* The first section of the part will fill evenly through the first contraction. However, when the second contraction and the rib are reached, the flow will become very turbulent and the pressure will drop. The pressure drop will cause surface and internal stresses to form that will cause the part to tend to warp.

- *Pressure will drop at the base of the rib.* The pressure may be too low in the rib area for the rib to be completely filled. The plastic will then have to be injected at a higher pressure and melt temperature in order to fill the mold.

- *Sharp edges in a part cause turbulence when the plastic flows.* Internal stresses occur at these corners. This stress makes the part very weak at this section. The solidification of the plastic is uneven, and high-stress lines that can produce cracks may form.

Modeling Techniques That Enhance Material Flow

Modeling techniques are available that can help the designer deal with turbulence in material flow. One solution is to add rounds to corners and other sharp edges, as shown in the following illustration.

Rounds on sharp edges.

Rounds make the flow of material into a rib smoother and less turbulent, therefore reducing the internal stresses that may cause the part to be weak. In addition, stress concentration areas are removed with the addition of rounds. Any external stress will be spread over the area and not located at one point.

One of the problems in the previous model was that the plastic could not get to the end of the part in the mold. The contractions, or thin walls, in the part stopped the plastic from reaching the end of the part. The designer can solve this problem by adding "flow runners." A flow runner is a short rib that gives the plastic more cross-sectional area to flow. Therefore, the plastic can get to other areas of the part without getting stopped in the contraction. The following illustration is an example of a flow runner.

Flow runner.

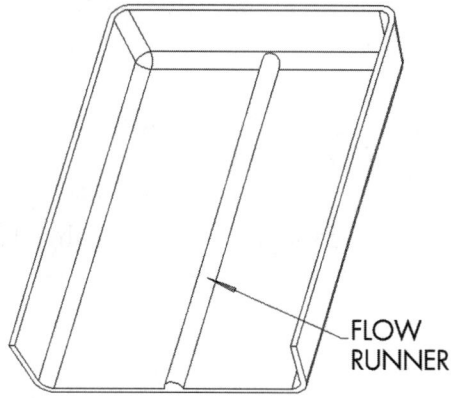

Another feature that cannot be ignored when designing parts is the tall, thin rib. You are dealing with a tall, thin rib when a rib is at least seven times taller that its width. Tall rib features have many applications, such as inhibiting electrostatic discharge, as described in Chapter 6.

Because these ribs are tall and thin, the flow path across them is short. The designer and mold maker must consider gate locations with respect to these ribs to ensure that the plastic will completely fill the ribs. With the ribs being tall, ejector pins are usually added to the mold to help aid in pushing the part from the mold.

Ejector pins might serve another function: they may be loose enough in the mold to allow the air to escape as the plastic flows into the tall ribs. This would be a source of air escape in addition to vents, as described in Chapter 1. Tall ribs usually have smooth surfaces, which allow a 0.5-degree draft, which helps the viscous material flow into them.

Managing Sink Marks, Flash, and Weld Lines

Undesirable sink marks, flash, and weld lines occur in plastic parts. These flaws are defined and discussed in the sections that follow, including methods of producing designs that eliminate or diminish these occurrences.

Sink Marks

Sink marks occur on the opposite sides of walls containing ribs and protrusions. The plastic flows into the mold at these locations and cools from the outside in. As plastic cools, it shrinks. More plastic at the base of the ribs and protrusions takes longer to cool and the plastic shrinks more. A sink mark can occur where the plastic is too thick.

Under the Info Measure selection of Pro/ENGINEER, you can determine the thickness of a section of the part. The Thickness command measures the thickness of a part and shows whether the part is thinner or thicker than the specified minimum and maximum values. If the area is larger than the maximum

Managing Sink Marks, Flash, and Weld Lines

value, the boundary displays a red color. If the area is smaller than the minimum value, the boundary is displayed in blue.

The plastics designer has the option of selecting which plane or planes need to be checked. This measuring tool lets the designer see which sections of the part may need to be modified to avoid sink marks. The following illustration shows the thickness of a part in cross section.

Cross section of a part showing its thickness.

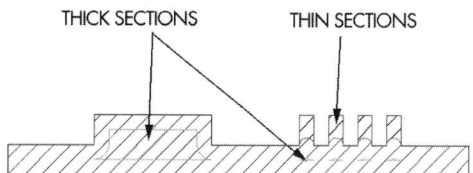

To avoid sink marks for ribs, a rib should be no more than 60 percent of the wall thickness. For thin wall parts, the rib may be the same thickness. For the gas assist injection-molding process, ribs can be any thickness. The reason is that the area between the rib and the wall is not solid plastic but is filled with a gas. Therefore, the real wall is much thinner than it appears from the outside.

Modifying the protrusion can solve the sink mark problem caused by protrusions, shown in the following illustration. The protrusion is changed to a partially empty protrusion, thus removing the excess plastic and the thick plastic area.

Sink marks and protrusions.

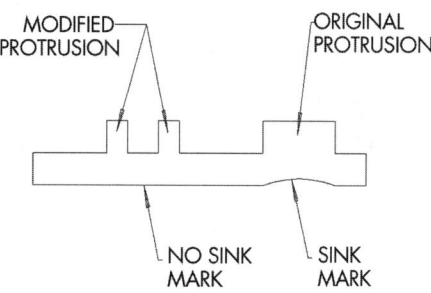

For bosses and tapers, the center of the feature should be hollowed out. The purpose of this is to remove the thick sections of plastic and to try to make the wall thickness even throughout the part. For example, assume you need a screw boss 40 mm tall, but the screw measures only 10 mm. Instead of making the hole in the center of the boss 11 mm deep, you should make the hole go down to 60 percent of the wall thickness at the base of the boss. This is shown in the following illustration.

Sink marks and bosses.

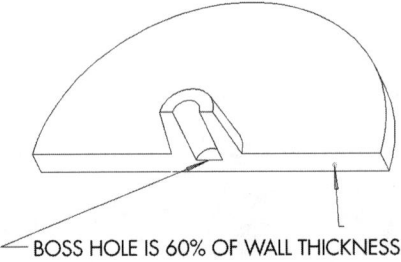

Sometimes bosses are very close to the edge of a part, as shown in the following illustration. The problem with this type of design is that a very thin piece of metal in the mold is used to make the space between the boss and the wall. This makes the mold very weak, and there is a good chance of this piece of metal breaking off. A better design is to make the hole open on one side of the metal piece that attaches the boss and the metal piece to the rest of the mold. The boss will function adequately, in most cases, depending on the manufacturer and the method of inserting the screws into the boss.

Improving a boss design.

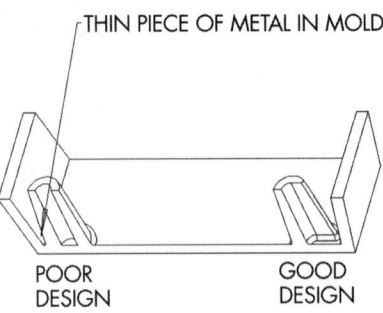

Managing Flash and Weld Lines

Flash is excess plastic that flows between core and cavity at the parting line, at inserts, at ejector pins, and at mold vents. It is normally about 0.025 mm thick. The designer can help prevent flash from occurring at the parting line by having core and cavity meet at a 5-degree angle from perpendicular to the parting line. Weld lines are lines that show up on a part where the flow paths of the plastic in a mold meet each other. The following illustration shows an example of this.

Weld line on a plastic part.

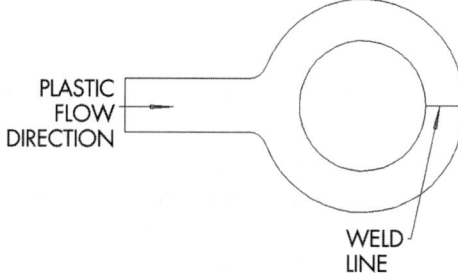

The plastic is injected on the larger side of the part. The plastic then flows down the part and splits in two directions. The plastic goes around the part and meets at the other end. Where the two plastic flow paths meet, a weld line occurs. The plastic is weaker at this weld line than at other locations in the part because there is no mixing of the plastic when the two flow fronts meet. The plastics cannot intermesh and form as good a bond as in the rest of the part.

The designer must decide if this weld line will cause problems in the functional aspect of the part. If it does, the designer can use Pro/ENGINEER features to, for example, change the location of the weld line. In this case, the designer might add flow runners to half the part to change the final location of the weld line.

Some parts require a mold that has more than one place (gate) into which plastic can be injected. This occurs when the flow path is too long for one gate. With two or more gates in a mold, weld lines will always occur. The designer must discuss the location of the gates with the manufacturer of the mold to ensure that the weld lines do not occur at high stress or thin sections in the part.

Inserted Areas

Sometimes it is more cost effective to mold textural information into a part than it is to add labels or print information on the part. However, for internationally sold products, this would be problematic (given language barriers) except for the existence of mold inserts. An insert is a piece of metal that can be placed inside a mold to replace a section of the mold. In the case of products sold abroad, a set of inserts containing the same information in various languages allows for the use of just one mold for a product. The original of this type of mold costs more than one without inserts, but the savings in not having to produce many molds can more than make up for the difference.

However, inserts have their down side. One problem with inserts is that the edge of an insert will not exactly meet the edge of the mold. It will not be perfectly flush with the mold and will create a fine flash or witness line. A good mold maker could make them flush but could not avoid the tiny amount of flash. The designer can user Pro/ENGINEER as an aid to making the insert invisible to the final purchaser of the product by incorporating a texture change or an elevation change. Often rounds or chamfers are added to this section to hide the witness line made by the insert. This way, the product is both attractive and functional.

Managing Tolerances and Relations

Maximum/Minimum Tolerance Modeling

A plastics designer must always be aware of the tolerances between features and parts that go together in assemblies. Pro/ENGINEER has a feature called Dim Bound (dimension boundary), under the SETUP menu, that allows the designer to examine the minimum and maximum dimensions of a feature. In the SETUP section of Pro/ENGINEER, you can change the feature to the minimum and maximum values (and values anywhere between) to determine if the feature is acceptable as a manufacturable item at a given value.

The designer can drive the dimensions to their minimum or maximum values by using the regeneration power of Pro/ENGINEER to determine the workable and manufacturable range of a product's potential volume. This type of testing is done on gears, for example, to ensure that the part will function properly within the manufacturing tolerance, as indicated in the following illustration.

Interference of gears.

Dimension Relations

One of the biggest advantages of using Pro/ENGINEER as your plastic design tool of choice is the fact that it is parametric. Features can be modified and features can be made into relations. For example, the following simple dimensional relationship corresponds to the part shown in the illustration that follows.

```
d1=d2+d3
```

A relational part.

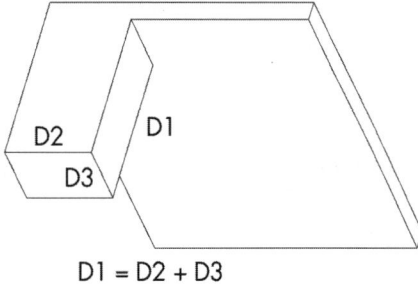

D1 = D2 + D3

The dimensions of the part shown in the previous illustration can be changed only relative to each other. This is called an equality relation. In plastic parts, the wall thickness at the base of the hole in the middle of a screw boss should always be 60 percent of the surrounding wall thickness. By using relations, this requirement can always be met, as represented by the following equation.

```
Boss_hole_thick = Wall_thickness * 0.6
```

The wall thickness of the screw boss in this example must be kept at 1.5 mm. Therefore, the hole in the screw boss can be made as a relation to the diameter of the boss, as represented by the following equation.

```
Boss_hole_dia = Boss_diameter - 3.0
```

The second basic type of relation is called a comparison relation. A relation may also be restricted to a certain length. This situation occurs, for example, when a part may not be closer than a specified dimension to avoid interference with another feature. The following illustration shows an example of this type of restriction. The concept is represented in the following relationship equation, where the total of the dimensions can never be greater than 10.

```
d1+ d2 + d3 < 10
```

Controlling the parametric model.

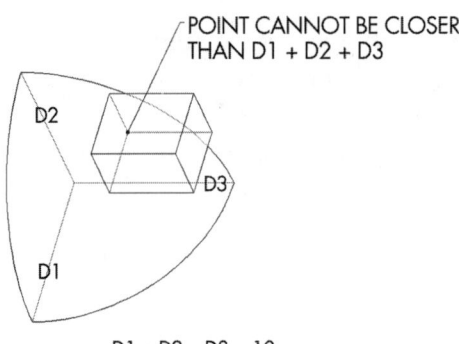

D1 + D2 + D3 < 10

Many options are available for the designer to make relations of dimensions. Most arithmetic, comparison, and mathematical functions are available to the Pro/ENGINEER user. Features in parts and assemblies can be made relational. Therefore, you can control which features of a part cannot be dimensioned in an assembly because the feature has already been made relational in part mode.

Pro/ENGINEER also has a function that allows the designer to make the feature read-only. This means that the feature cannot be changed and does not get regenerated. However, you may add features to a read-only part.

Evaluating Features

✓ **TIP:** *The read-only feature is useful when you are working on large-featured parts and you are near the end of model creation. The model regenerates very quickly because all features up to the read-only part are not regenerated. The problem with the read-only feature is that all features before it cannot be changed. Therefore, if you want to modify the previous features you must turn the read-only feature off.*

Evaluating Features

Features can be created that look adequate but are not manufacturable. An example would be a feature that has no draft at the parting line. The part is not manufacturable. A possible solution to the problem is to use the Graph function in Pro/ENGINEER. This function allows you to create a relation or curve that defines the shape the final curve requires if it is to be manufacturable. The first of the following illustrations shows a sample Graph file. The second of the following illustrations shows a part with no draft at the parting line.

A sample Graph file.

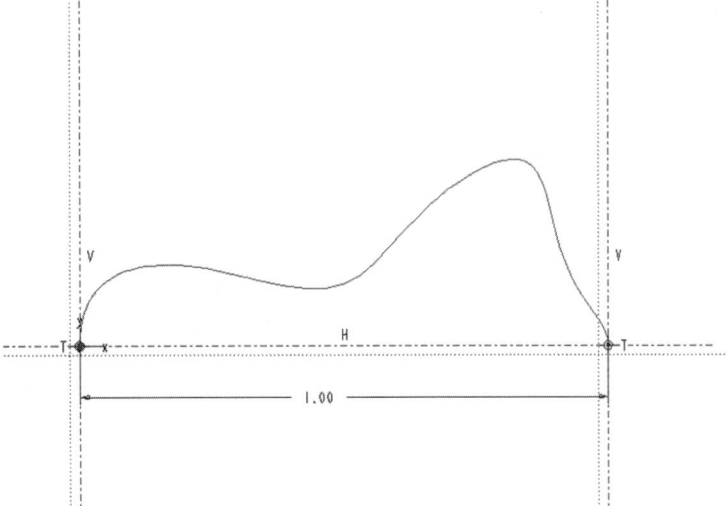

Part with no draft at the parting line.

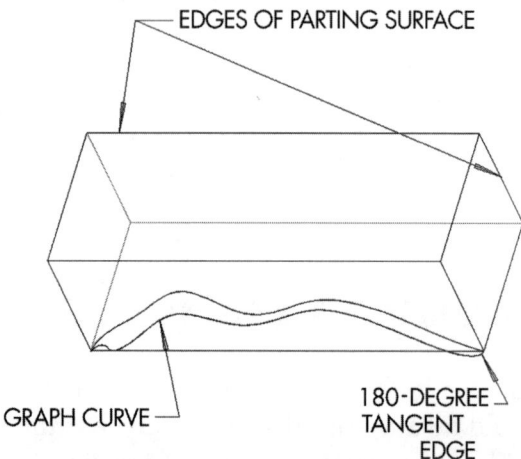

The Graph function uses a curve, line, or spline that describes a desired shape. The Graph function is found under the Datum section of Pro/ENGINEER. The curve is created and named, but is never seen on the part. You can then use the Graph function to redefine or replace a feature on the part. Having center lines in the graph function can make tangencies to existing curves in the base part. An example of a part modified with the Graph function is shown in the following illustration.

Part modified by the Graph function.

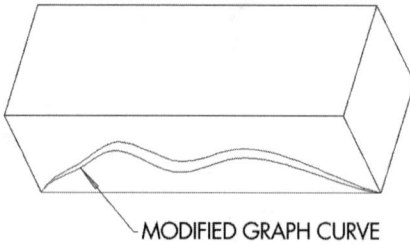

Summary

Parts intended for production must be designed for manufacturability. The ideas, techniques, and process aids discussed in this chapter are intended to help you produce marketable and usable products. Keep in mind the following design rules and suggestions.

- Avoid sharp corners by adding rounded features.
- Use smooth walls wherever possible.
- Be aware of trapped air and try to avoid this situation.
- Add flow runners to enhance material flow.
- Try to avoid tall ribs if possible, they are always difficult to fill with plastic.
- Keep the rib thickness to no more than 60 percent of the wall thickness.
- Keep the wall thickness as uniform as possible to avoid sink marks.
- Avoid putting features too close to the edge of walls. They may cause the mold to have thin pieces of metal that may break.
- Keep weld lines away from high stress areas.
- Use inserts for products that change only in one location in a mold. They are less expensive than entirely new molds.
- Use relations when parameters are useful in modifying designs.
- The Graph function can be used to describe a specific shape from one edge to another.

This chapter has also discussed the use of Pro/ENGINEER and its tools for making manufacturable parts. The next step is to apply the tools to the design of plastic parts, the subject of Part III.

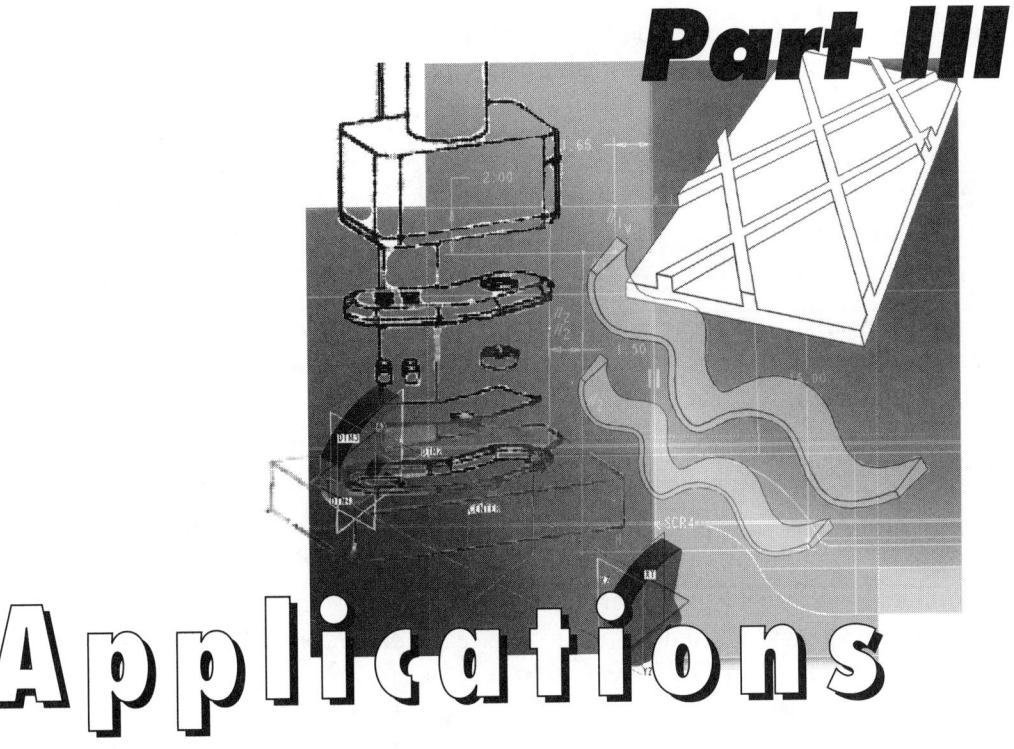

Part III

Applications

Hands-on Design Projects

Part III is devoted to putting the discussion to this point into practice. The project examples and exercises (as chapters in total and within chapters) presented are representative of the application of everyday plastic design tasks, using Pro/ENGINEER as the modeling tool of choice. The exercises focus on the detail design of plastic products and components. Emphasis is placed on manufacturing issues likely to arise during the project design cycle.

The examples have been chosen from a variety of plastics technologies, from injection molding to blow molding. The object of the project designs chosen is to show you how to create models representative of what you would deliver to the toolmaker. All of the examples are centered on techniques for mass production of the design components inherent to the representative projects.

The project deliverable in each example is intended to be a 3D model ready to be delivered to the toolmaker for manufacture of the mass-production tooling. To produce the tooling, the 3D file would suffice as a stand-alone deliverable. Critical dimension drawings could be made to clarify design intent and identify any areas of special interest to the designer.

In a professional manner, each of the components designed will start with a part start program, which will provide the default datum planes coordinate system, default views, and an accuracy of .0003. You may, however, vary the accuracy where the need arises. Consider the accuracy range of .0003 to .0008 to be appropriate due to the complex nature of some of the geometry.

Each exercise endeavors to follow a step-by-step path to solving the modeling or design problem given. Some of the examples are based on product design techniques, and the remainder on component design techniques. The actual geometry dimensions used to create the features are given where they are relevant to the discussion. However, where you are able to follow along, creating something similar on your screen, the details of dimensions are avoided to provide for flexibility.

With this approach, if, for example, you are building a box and prefer to build it in inches instead of millimeters, the overall lesson will not be harmed. It is more important that you master the technique than follow a dimension-by-dimension series of steps by rote. As plastic material technology and component processing advance at a rapid rate, it is advisable that early in your modeling task you work with the resin suppliers and toolmakers to stay on top of what the latest advances are in your area of technology. Part III begins with the design of a compact disc player intended for the injection molding process.

Chapter 8

Portable Compact Disc Player

An Injection Molding Product Design Exercise

Introduction

The discussion in this chapter focuses on the approach and the techniques involved in designing plastic housings for a compact disc (CD) player. Step-by-step detail covers the design for the external housings from concept to finished individual component files. The technology chosen for this project is plastic injection molding because of the mass production nature of the design and the high tolerance demands of the product's finished components.

Pro/ENGINEER offers users a number of approach paths to model a product such as this CD player. One approach is not necessarily better than another. The approach chosen in this project would represent a single designer working on the original product layout to describe the bulk of the external geometry prior to splitting the product into parts.

Although tools within Pro/ENGINEER allow for the creation of complex geometry in a relatively straightforward method, the geometry must be manufacturable. The project example chosen follows techniques that are robust when modeling plastic components. Another key learning feature of the project is that the line of draw (parting direction) of the molded parts must be maintained during the modeling process if a manufacturable design is to be achieved.

Project Description

You are required to model a new CD player using an existing internal chassis and drive motor used in another product. The assembly has a top housing, a bottom housing, and a flip-up door on top of the unit. There will be a battery box on the bottom cover, which will allow access to the batteries. There are four side buttons and a jack outlet also to be designed, which work with the existing internal components.

The styling department has completed its design, considering the size of the internal geometry. You have been asked to keep the overall size of the model flexible for fine-tuning in anticipation of minor changes.

Due to the nature of the geometry created in this project, it would be best to start with a look at the finished product, shown in the following illustration. Take note of the flip-up door on the top of the set and the location of the main parting plane where the top and bottom housings are to join.

Product design of CD player to be modeled.

Applying Design Philosophy

The product development cycle for modeling the individual components of the CD player needs to be considered in the scope of the overall design and its characteristics. It could be said that the components dimensionally rely on one another. This is obvious where the top door is designed to match the top housing, and where the battery outlet is designed to mate with the bottom housing. This product-modeling project may be approached as a task in which the external geometry is modeled and the components cut out of the geometry.

Pro/ENGINEER offers users a number of techniques that would enable them to tackle this design task. One could use the internal existing geometry to act as a guide to modeling the external geometry by means of designing within an assembly and applying the features to individual components. A skeleton model could be created and used in an assembly to aid in modeling the component geometry. This type of project may also be tackled by using the master part file or master model technique. Use of this technique would make all of the components dependent on the master part.

There are a number of issues predefined in this product design that could be modeled into a single component, with the piece parts split off from the single part file. There are also three techniques for applying this concept. A master model part file could be established and the components cut out of it, or a skeleton model could be used to model the common geometry where the geometry would be handed off to the individual component files from the skeleton model. This would result in a modifiable component, with changes flowing through any of the components affected by the changed geometry. This technique would make all of the components dependent on the master part file for their overall geometry.

A third approach would be to start with a reasonably intelligent start part and model the components using that start part as a guide. The components would all have internal to them the same geometry as the original start part, and would remain as stand-alone files. The start part model approach is used for this project.

> **NOTE:** *Subsequent chapters will present other design approaches.*

Looking at the modeling requirements enables you to move on to the modeling plan for the project. The first step in this design is to identify what the known characteristics of the product are. Making a list of them will help to identify any special considerations to be included in the modeling plan.

- The parting surface for the overall design is a flat plane that divides the top and bottom housings.
- The external profile of the geometry has been established by the styling department.
- The assembly will be held together using four screws.
- Texture will be applied to the external surfaces, drafted at 4 degrees.
- The flip-up door is modeled as a match to the top housing geometry.
- The battery door is modeled as a match to the bottom housing geometry.
- There are four buttons centered about the parting line. Therefore, the button geometry will affect more than one part.
- The headphone jack is located on the side of the unit.
- The internal chassis exists in another product and the geometry is available.

The previous list of data in conjunction with the styling plan drawing enables you to begin the modeling plan for the product. The order in which you approach the piece part design will save you valuable time. For example, there would be no point in trying to model the top door until the geometry for the top housing was complete. The bottom housing shares geometry with the battery door, and therefore also falls into this category.

If the internal existing geometry were modeled into a start part, that part could be used as a guide for all of the other components. The existing geometry is in a fixed position and critical items (such as hold-down locations and switch positions) could be highlighted to aid in the overall component design. The plastic housing design would be an easier task if the existing geometry could be referenced during the design cycle. Assembly of the components would be a simple match of the coordinate systems on each of the components. Because they were all started from the same start part, a match is guaranteed.

Modeling Plan and Process

To start the modeling plan for this design, consider the technique chosen. For this example, you start with an intelligent start part and build the component structure off that part file. To model the information in the start part as solid geometry would defeat the purpose of the start part. It is not intended to be a solid model structure for the known geometry. It is only a group of reference data that will enable you to quickly model the product geometry. With this in mind, consider the start part itself as a beginning to the modeling plan and expand to include how the component geometry is derived from using the start part.

It has been established that there are some features of the design that are in fixed positions due to the reuse of an assembly from another product. There are other issues that also contribute to the overall design that are known at this time. In considering the modeling plan for the overall design, you must first consider the components or features of the start part. What should the start part have in it based on the known information at hand? Listing the attributes of the part may help to identify what needs to be modeled.

- Start with a generic start part to get the default datums, planes, coordinate system, default views, and part accuracy setting.
- The external geometry may be produced to represent the chassis assembly in simplified form using simple surface structures.
- The compact disc itself may be modeled as a reference part.
- The hold-down location where the internal chassis will clamp to the plastic housings may be identified and modeled, using a surface.
- The screw boss locations may be identified by using datum points.
- The button and jack locations and sizes may be represented with datum curves created on a datum plane.
- The batteries may be modeled in position as surface geometry to act as a space claim.
- The flip-up door pivot location may be modeled as a point and axis.

- The profile outlines for the product external shape, as well as the profiles for the top door and battery cover, may be modeled as datum curves.

It is important to note at this time that the information placed into the start part model will affect the geometry of more than one component. Therefore, the information may be considered as representing root features for the entire product design. Model geometry for the components will reference this geometry. Because the geometry modeled into the start part are all datum surfaces, curves, and points, these entities may be placed on layers and blanked, leaving only the geometry for the individual component showing. Once the start part has been created, it is easier to complete the modeling plan for the overall design. The following illustration shows the start part with the root feature information labeled.

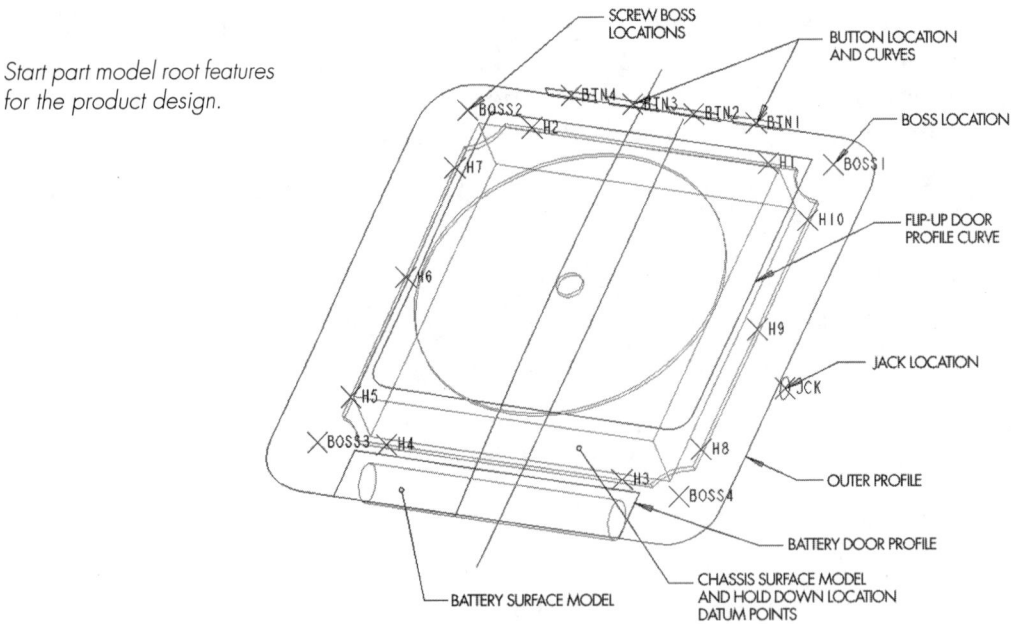

Start part model root features for the product design.

The modeling plan for the component structure and design intent may proceed based on using the known attributes of the start part as root features. The original design intent of the product as shown in the overall assembly can help to identify what features have to be considered from a styling point of view. Tak-

Modeling Plan and Process

ing the original illustration and labeling it with the styling information will enable you to complete the identification of the modeling requirements.

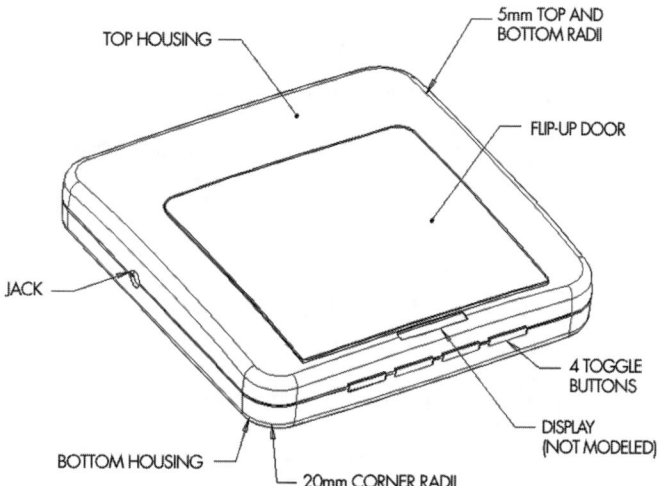

Styling features labeled.

The intent of this styling illustration is to identify what sizes the cosmetic features are. There are number of issues to consider for the development of the modeling plan. The modeling plan may be considered a list of wishes, as it may not be possible to fulfill all of the requirements. An example of this is that a round of a value of 10 mm was identified for the top radius all around the top housing. This size may simply not fit when the design of the component in reference to the internal geometry is considered.

Intelligent Start Part Plan

The plan for the modeling process may include all of the issues or just the most significant ones, and the plan may exist in more than one form. You may have a product modeling plan and a component modeling plan. They may be written down, or just a strategy you have come up with while spending an hour on the freeway. In any instance there is no possibility of having a plan unless you are aware of what you are modeling. In cases for which most of the issues are unknown, the modeling plan or strategy and its maintenance may evolve as you design a product.

Based on the known information for the CD player, you can begin to identify a strategy. Listing the strategy as modeling steps helps produce a plan with definite, albeit flexible, structure.

- Create the start part with the root features in it.
- Model the external geometry of the product.
- Split the design into a top and bottom housing.
- Shell the geometry and place the interlock lip in position.
- From the individual top and bottom housings, cut the battery cover basic geometry and the flip-up door basic geometry as separate components.
- Assemble the components to keep a watch on the overall design.
- Complete the detail design for each component.

Armed with this information and an approach, you can begin modeling the CD player. The section that follows provides a phase-by-phase and step-by-step walkthrough for creating the external geometry of the product.

Create Product External Geometry

EXERCISE

Phase 1: Create Start Part

When creating a start part to be used as a common component in the product design, all of the generic information should be included in it. The root features modeled as surfaces, planes, curves, and points do not take a long time to regenerate and may be turned off via the layer button if these features are placed on layers.

The most relevant starting point for this modeling job would be to pick up the generic start part used when starting a new component and add the root features to that model. By doing this, all components you are designing will have the same view names, accuracy setting, and default datums. An accuracy setting of .0003 to .0008 is recommended due to the complex nature of the finishing geometry. Tiny rounds can contain very complex surface structures, and draft can cause a number of tiny edge geometry check problems. The steps for phase 1 follow.

1. Call up the generic start part. Rename the datum planes to names that will make sense in your design. On the parting datum plane, create datum curves representing the external profile, flip-up door profile, and battery door profile. On the center datum plane, place datum curves representing the overall height of the product. The result of this step should be the basic overall sizes of the part modeled using datum curves, as shown in the following illustration.

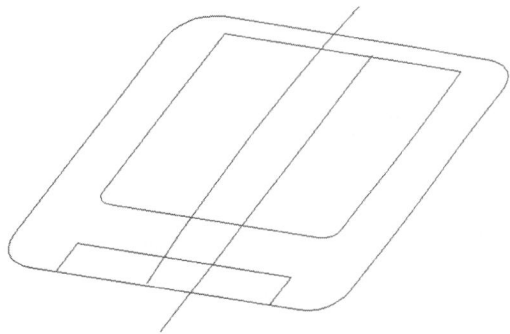

Root features representing the overall size created using datum curves.

2. Establish the mechanism used to represent the chassis assembly. It is unnecessary to completely model the chassis assembly for this task. Wherever possible, the important features (such as hold-down ribs) should be identified where they mate with one or the other housing. The plastic housings will be used to hold the chassis in place in the assembly. The plastic housing must have clearance inside the assembly where there is not a requirement to clamp the chassis in place. Therefore, the chassis model should include a basic representation on the assembly in block form, as well as the intended loca-

tions of the hold-down ribs. The rib locations may be transferred to the housings further into the modeling cycle. It is advisable to represent the chassis geometry using surfaces to enable you to check for clearance or interference during the detail design of the components.

3. Create an extruded surface with closed end patches representing the chassis overall dimensions in the desired location. The button reference will provide the height location of the chassis surface structure because the buttons are defined as being on the parting plane of the housings. Create datum points along the outside edge of the surface structure where hold-down ribs from the housings may apply pressure on the chassis without damaging the assembly, as shown in the following illustration.

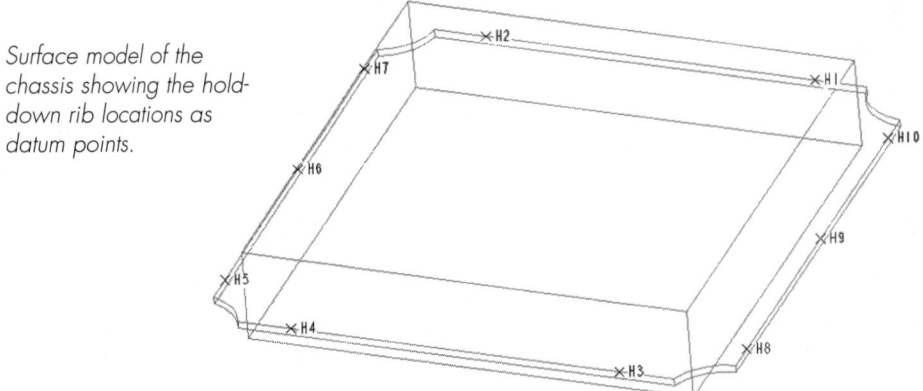

Surface model of the chassis showing the hold-down rib locations as datum points.

4. Model the location of the batteries as a surface model to make a space claim in the housings for the batteries. The screw boss locations may be defined at the same time to also make sure there is space enough for them in the assembly, as shown in the following illustration.

Modeling Plan and Process

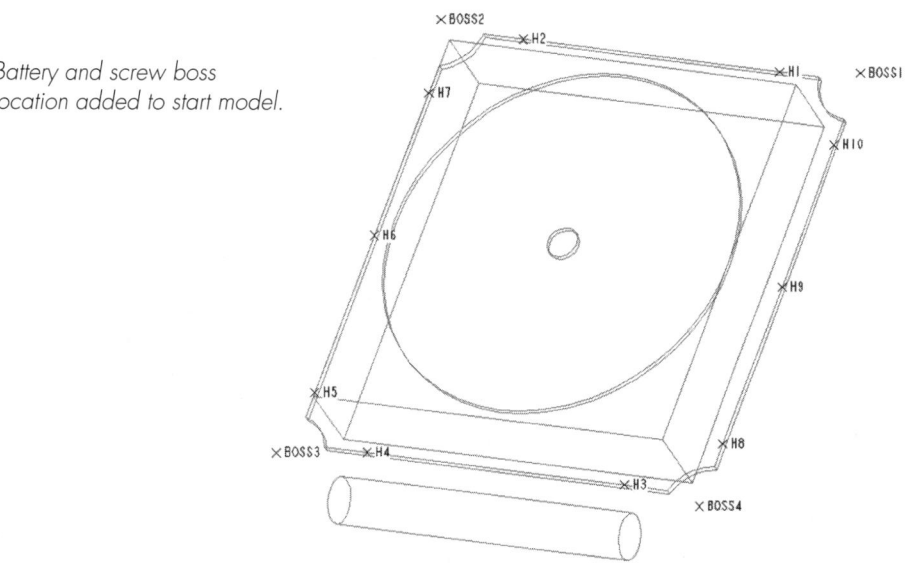

Battery and screw boss location added to start model.

5. Model the size and location of the buttons (keys) and jack. The button locations are referenced to the chassis, and the jack is located along the side of the chassis. These items may be modeled as datum points, representing the center point of the item, and as datum curves representing the external profile. If there were a need to modify the button location during the detail design, the only thing to modify would be the center point location of the button. The datum point would act as the root feature for all of the button geometry in this situation. The key and jack additions are shown in the following illustration.

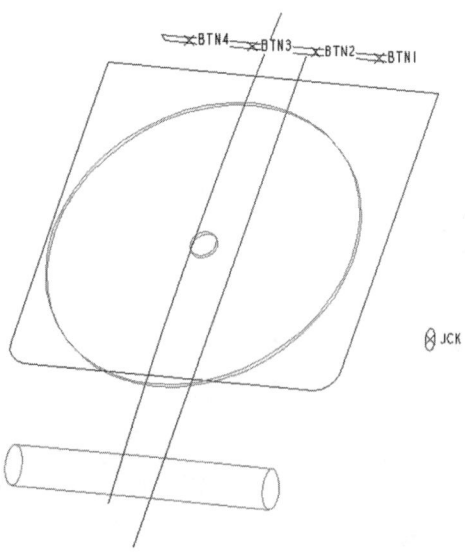

Key and jack locations added, completing the start part design file.

Phase 2:
Model Top and Bottom Housing Common Geometry

The start part may be used as a starting position for any of the piece parts in the design. The buttons for keys are an example of components for which you could use the common start part model. In this phase, a copy of the start part model will be made and the design of the product will proceed, wherein you will add the overall solid model geometry for the exterior of the product. Because the top and bottom housing must mate, what better way to ensure they do than model the geometry simultaneously?

The external styling features will also be added to the design to finalize the outside shape. At this point in the design, the model will begin to take the appearance of a manufacturable product. Consideration will be given to the external surfaces needing draft angles for texture and the external cosmetic rounds. The steps for phase 2 follow.

1. Create a protrusion extruded from both sides of the parting datum plane, referencing the external profile of the product. This results in a simple block extending from both sides of the plane. Use a value greater than the overall height to allow for trimming of the top and bottom surfaces.

2. Create a cut following the top datum curve representing the crowned surface of the product. This cut is an advanced variable section sweep using the parting plane as the pivot plane. The datum curve representing the overall height is used as the trajectory. The pivot direction plane will keep the section perpendicular to itself during the feature creation. This technique, commonly used in plastic part creation, will help keep the product manufacturable. The cut is shown in the following illustration.

Cut added to crown the top of the product.

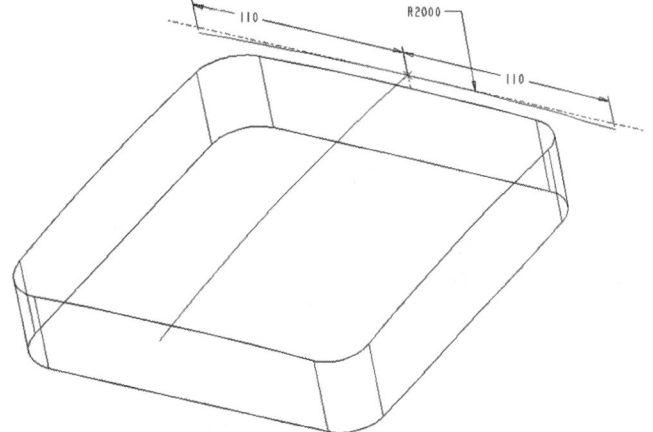

3. This process may be repeated for the bottom crown on the product. It is advisable to create these features sequentially so that they are relatively easy to find should you need to change the profile of either of the features. Crowning of a large plastic surface is a good idea because it helps prevent oil canning (wavy) or sagging areas on surfaces during molding of components or parts. In addition, a

more cosmetically pleasing design will result if large, flat surfaces are crowned, as shown in the following illustration.

Resulting block with the top and bottom crown added.

Phase 3: Add Draft and Cosmetic Features

Draft in this product design must reference the parting plane or surface and be applied to surfaces in the correct direction on the top and bottom housings to enable the product to be manufactured. The exception to this rule would be places where slides were added to the mold and the geometry would need to refer to the pull direction of the slide for the drafting plane to create the draft. In this example, the draft is central about the parting plane.

1. The draft feature for the external walls may be created in a single feature using the split draft option in the menus. The first step in the feature creation process is to select the surfaces to which the draft will be applied. The second step is to identify the plane where the draft split will reference. The final pick is to identify the plane from which the draft will be measured. From this point you enter the values for the draft in the appropriate direction. The draft is created such that the largest part of the geometry is at the parting surface, with minus draft from that point.

Modeling Plan and Process

2. With the draft in place on the exterior walls, the rounds may be added to the top and bottom corners of the product all around. This is most easily accomplished by placing a simple round following a tangent edge all around the top and bottom of the product. This round should be created independently on the top and bottom edges (two features) to allow for the possibility that change will occur in the piece parts. Because the product shape will be split into housing parts at the end of this step, it is of value to the overall design to keep these cosmetic features independent of one another. The external draft and rounds are shown in the following illustration.

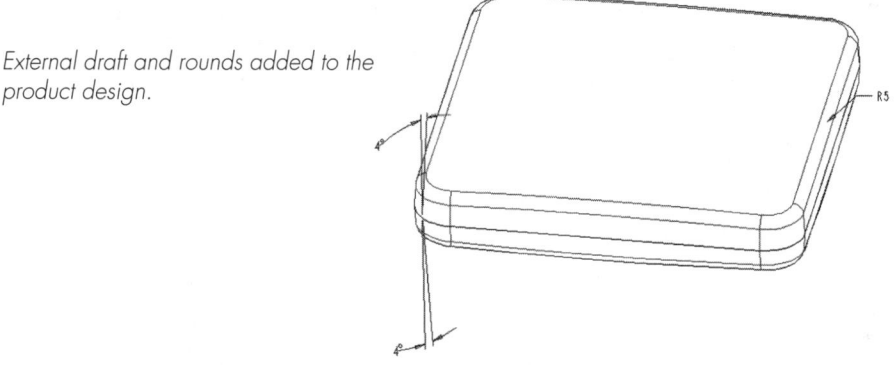

External draft and rounds added to the product design.

Phase 4: Create Top and Bottom Housings

The top and bottom housings may be created from the product shape at this point in the modeling cycle. The existing part should be copied to another name for use as the bottom housing, and in turn, this component may be renamed for use as the top housing. Starting with the top housing, you will begin to form the independent halves of the top housing component.

The top and bottom housings are intended to mate at the parting plane. If the surfaces of the housings were to be mated, there would likely be some sort of mismatch due to manufacturing tolerance and warpage on the fit between the two large flat surfaces where the components mate. This could leave your design looking cosmetically unprofessional.

If you provide a shadow groove all around where the components meet, it would help to disguise any mismatch in the flatness or the profile of the two components. This feature is a 1-mm groove that runs all around the exterior of the parts once they are assembled. The actual creation of the interlock features will provide for the 1-mm space all around. The important thing at this point is to allow space for the features. It is necessary therefore to create the cut for the top housing 0.5 mm from the top housing, and to repeat the same operation for the bottom housing, making the 1-mm space in the interlock lip area. The steps for phase 4 follow.

1. Create a cut, using the center datum plane as the sketching plane, 0.5 mm offset from the location of the parting direction plane. The cut should be created on both sides of the plane to completely remove the bottom housing from the part.

2. Create a shell for hollowing out the solid geometry at a wall thickness of 2 mm. This feature will complete this step in the geometry creation of the top housing.

3. Repeat steps 1 and 2 for the bottom housing part. Both components have been shelled out to a constant wall thickness and the interlock lip may proceed. The cut and shelled housings are shown in the following illustration.

Modeling Plan and Process

Top housing shown cut and shelled to a constant wall thickness of 2 mm.

Phase 5: Create Interlock Detail

A mechanism must exist for sealing the housings against each other with 1-mm spacing at the position shown in the following illustration. The lip feature may be used to create the seal between the two components, which will provide a suitable clamping surface and help to align and position the housings during the assembly process. The lip feature is created using the Tweak command. It is a simple process of offsetting a surface at a specific distance from the edge of that surface.

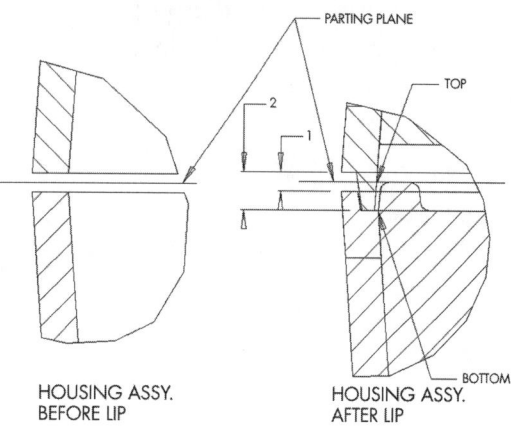

Lip interlock features for top and bottom housings.

An added bonus to using this command is that you are able to specify an angle on the surface edge of the lip, resulting in a self-aligning mechanism when used in this application. The male portion of the lip is applied to one housing and the female to the opposite housing. The resultant geometry is shown in the previous illustration.

This feature is created on each part such that it applies all around the perimeter of the components. To create the feature as a male lip, use a positive value for the offset distance. A value of 1 mm is added to the male lip to close the seal all around the part, making up for the 1-mm spacing from the previous step. The female lip is created using a negative value for the offset distance. The creation of the male interlock works as follows.

1. Using the Tweak Lip command, select the edge surface on which to apply the lip.

2. The software will then ask for a reference edge. This will identify to the software which edge you wish to measure the offset distance from. This edge may be selected as a loop to obtain the tangent edge chain in a single pick.

3. The next requests from the software will be the surface to be offset, the offset distance, and the angle for the outer edge surface of the feature. For the purpose of this model, select 2.5 mm as the offset distance and 10 degrees as the angle. The following illustration shows a before-and-after diagram of the lip selection process.

Modeling Plan and Process

Lip selection process.

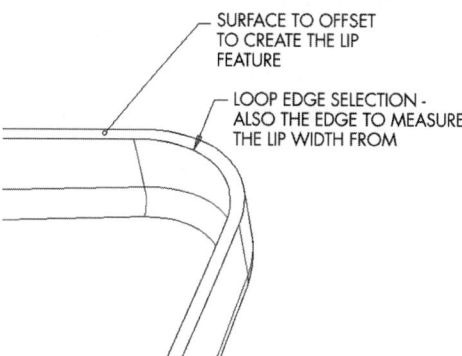

4. Repeat the process for the bottom housing using a negative value of −1 mm for the offset distance. This will complete the interlock lip step.

With the interlock lip applied to both, the housings may be further developed. At this point in the modeling cycle, the top and bottom components are independent but are also destined to be used at root parts for other components. The sections that follow continue development of the top housing and flip-up door. The steps that apply to both housings are provided.

Phase 6: Create the Flip-up Door

Using the top housing component, create another copy of the part called Flip-up Door using the Save As menu pick. The basic geometry for the flip-up door will be created referencing the datum curves modeled in the start part. The steps for phase 6 follow.

1. Call the flip-up door to the screen and turn on the layers where the start part geometry resides.

2. Create a cut from the parting datum plane using a sketched section. To create the sketch, use edge on the datum curves representing the flip-up door outline. By removing all of the external solid geometry, you are left with the beginning structure for the door.

3. File this component, to be completed later.

The first of the following illustrations shows the flip-up door sketch. The second of the following illustrations shows the basic geometry for the door.

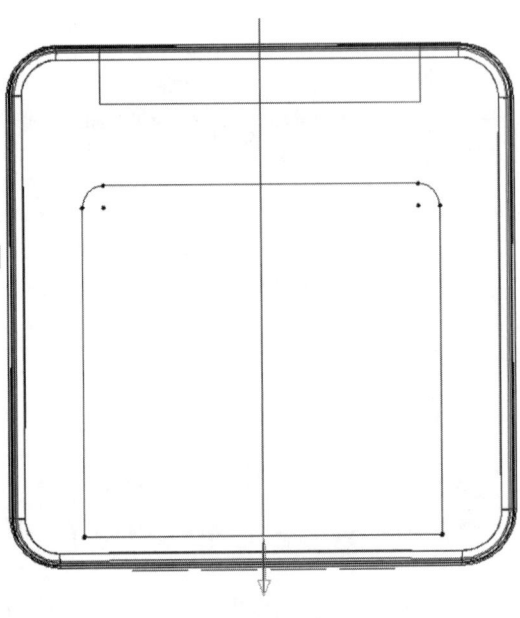

Sketch used to create the flip-up door. (Bottom arrow points to the material to be removed.)

Basic door geometry created.

The flip-up door has now been created in its most primitive form. Many previous features of the modeling process have become root features for the door. The door geometry—including pivot features, latch, and interlock—may be completed at a later time. The top housing component design resumes with phase 7.

Phase 7:
Create Top Housing Door and Support Features

You now need to cut out the top housing flip-up door opening. As in the previous phase, a cut will be used to create the door opening referencing the start part datum curves. Because the door must have clearance to be assembled, an offset sketched entity will be used to offset the datum curve geometry to allow for the assembly clearance. The features relating to the door cut will also be added to keep the door detail sequential in the model tree. The steps for phase 7 follow. The illustration that follows the steps shows the cut, draft, and chamfer added to the model.

1. Create a cut, referencing the parting datum plane, for removing the material for the door opening. This cut is perpendicular to the parting datum plane and offset from the datum curve by 0.3 mm. Apply a round of 3 mm to the two square corners of the door opening. This feature will allow for clearance all around the door opening.

2. Create curve-driven draft all around the door opening on the internal surfaces of the hole. A value of 1 degree will be acceptable for the draft because the surfaces will be polished.

3. A finishing chamfer all around the opening will help to disguise mismatch between the door and the housing when assembled. Create a chamfer using the d1xd2 (distance along a surface) option on the tangent edge all around the door opening. The value for the chamfer should be 0.5 x 0.5 mm.

Cut for door, round, draft, and chamfer added to model, with one corner shown.

Phase 8: Add Door Finishing Features

A lip or stop must be added to stop the door from going through the housing when it is closed. The cut for the door has left an edge where a lip or stop may be created to facilitate this requirement. To add a protrusion all around the profile of the door is not easily done with a standard protrusion. An advanced variable-section swept protrusion would be able to travel all around the profile while staying perpendicular with the parting plane. The steps for phase 8 follow.

1. Create an advanced variable-section sweep, with the pivot direction set to be the parting plane along the bottom edge of the door cut-out. The sketch for the section may include the draft angle, or you may add the draft as a separate feature.

Modeling Plan and Process

2. Create the section for this feature to slightly bury the feature into the existing solid geometry. This is done to ensure that as the section travels around the selected profile it does not slightly lift off the existing solid. This can occasionally happen, depending on the curvature of the edges you are following. An advanced variable-section swept protrusion will take a single section profile and follow one or more trajectories with the section profile.

The first of the following illustrations shows the section used for the swept protrusion. The second of the following illustrations shows the finished door stop feature.

Section used for the door stop feature.

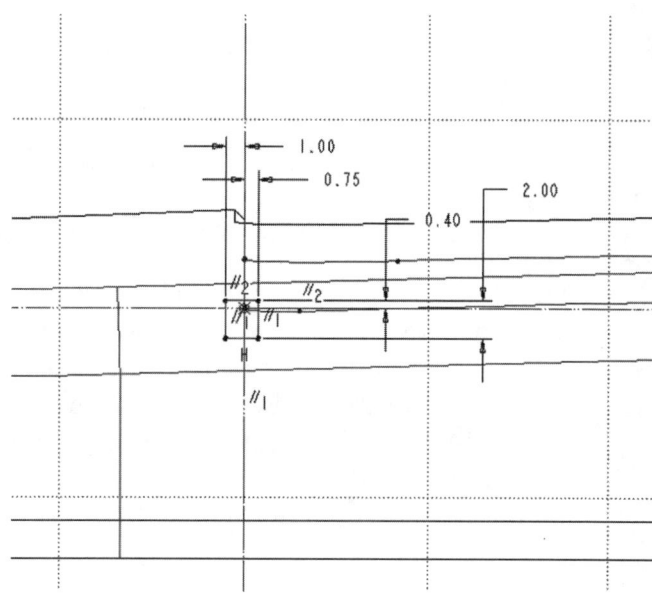

Finished door stop feature.

With the door stop feature in place, phase 9 returns to holding the structure together.

Phase 9: Clamp Inner Chassis and Secure Interlock

The inner chassis needs to be clamped into position using ribs from the top and bottom housings. The positioning of the hold-down ribs has been defined in the start part geometry as datum points on surfaces. For the purpose of this phase, consider the geometry as it applies to one of the hold-down ribs. Look at a cross section of the assembly and determine what the design intent of the hold-down ribs is. The following illustration shows the position and purpose of the hold-down ribs.

Position and purpose of the hold-down ribs.

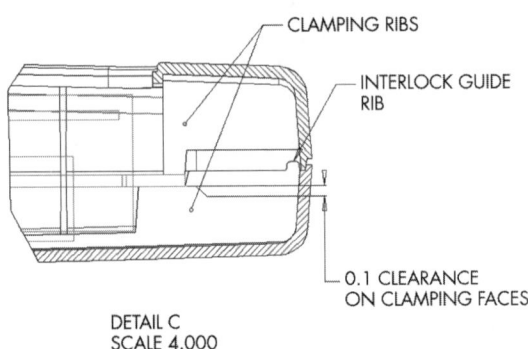

The previous illustration identifies the location where the hold-down ribs will apply. Another function of the hold-down ribs will be to stop the chassis from moving from side to side internally in the assembly. As a bonus, the ribs on the bottom housing, if extended to the exterior of the wall, may be used to secure the interlock lip into position and improve the integrity of the assembled unit. If the interlock lip is locked into position, the side walls will resist being pushed in and unseated from their location. This feature stiffens the structure of the entire assembly, and helps control the mismatch in the external profile.

For modeling purposes, the assembly order dictates that the housings be clamped together, then the chassis clamped into position prior to the screw bosses being seated during the assembly of the screws. The housings have already been modeled as touching. Therefore, even though you are creating clamping features, they will be offset from the surface by 0.1 mm to ensure that they are not touching before the housings touch. In the following steps for phase 9, the rib example will be applied to the lower housing to encompass the interlock alignment rib.

1. Create a protrusion from both sides of a datum plane through the point labeled H9. The protrusion should be created all across the component but have clamping sections included in the geometry. The rib thickness should be 1 mm to allow for draft to be added without exceeding the 60-percent thickness rule discussed in Chapter 6. The following illustration shows a 3D view of the housing containing ribs.

3D view of the housing with the ribs included.

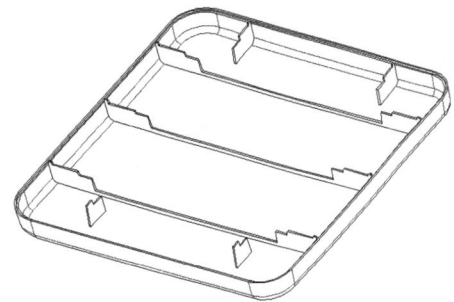

2. Draft may be added to the external surfaces of the rib to enable manufacture of the component. The draft may be modeled at 1 degree unless the wall thickness at the bottom of the rib exceeds the 60-percent rule. If this is the situation, the draft may be reduced to 0.5 degrees. This is typical on tall ribs. The draft feature may be created referencing the hold-down pads to guarantee that a width of 1 mm is maintained on the clamping face.

3. Repeat the previous for the rest of the hold-down ribs in both the top and bottom housings. Interlock alignment ribs may be added to ensure that the external walls are supported where there is an unprotected area of the component. It is difficult in this case to determine how close together to place ribs for aligning the external walls. For a reasonably flexible material, the ribs would be closer together than would be required on a stiffer type of plastic. Experimentation may be necessary to arrive at the optimum number of ribs. However, because steel would need to be added to the mold (welding) in order to remove a rib, ribs are more easily added than removed. Fillet radii at a value of 0.5 mm are needed at the base of the rib detail to add strength to the rib detail and to aid material flow.

4. Repeat the hold-down rib modeling process for the top housing. The following illustration shows the top housing with ribs in place.

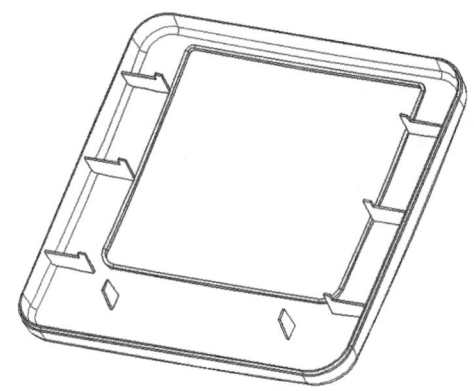

Top housing ribs added.

The features for the modeling of the ribs have been kept sequential. If you decide to delete a rib, you may delete that rib and its associated draft without destroying other features or references. In cases where ribs are the same and at different locations in the part, the ribs may be patterned or copied to the new locations. If the ribs are patterned, it is advisable to use the Pattern menu pick under the Group function to pattern the feature. Using this pattern allows for the removal of one of the pattern instances if desired further into the modeling cycle.

Phase 10: Add Screw Bosses

The screw bosses may be added to the assembly such that there is a 0.2-mm clearance between their respective mating surfaces. This allows the screw boss faces to be the last thing to come together when the assembly is clamped together via the screws. This feature will allow both the clamping of the internal features and the housings to be seated in the correct position. To help avoid sink marks, the wall thickness of the screw boss geometry follows the same 60-percent rule as the rib geometry. The cross section of the finished screw boss in the following illustration indicates the modeling required. The list that follows specifies details of the cross section. The steps for phase 10 follow the list.

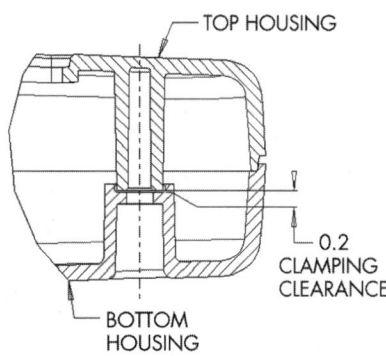

Screw boss cross section.

- The screw boss mating surface is inside the bottom housing to allow for a reasonable length of screw.
- The bottom housing has a ring on the boss to capture the protrusion from the top housing.
- A 1-degree draft is applied to all surfaces. The screw hole may be 0.5 degrees to minimize the wall thickness at the bottom of the screw hole.
- Radii are applied to the sharp edges (where stress may be a factor) to add strength to the plastic and help material flow easily.
- The depth of the screw hole follows the 60-percent wall thickness rule to prevent a sink mark in the plastic housing.
- To model the screw boss, start with a protrusion 6 mm in diameter, created from an offset datum plane, to derive a cylinder.

1. Revolve a cut about the cylinder axis to form the internal geometry.
2. Add draft, chamfers, and radii.
3. Copy or pattern the features to the other three locations. The following illustration shows the top housing screw bosses.

Top housing screw bosses added.

It is important to note that where possible the screw bosses should not be attached to the vertical outer wall of the housing. If you must locate a screw boss on the external wall of the housing, you need to be aware that the technique of using the screws to sequentially clamp the assembly together may not be possible. This is because the interlock lip has been pulled together close to the screw, preventing any further clamping. By having the screw boss located a distance away from the wall, you have ensured that if the components are manufactured correctly the design intent will be achieved. The remainder of the product may be modeled using the techniques shown in previous phases.

Conclusions: Intelligent Start Part

The example in this chapter referenced designing a plastic product using common data relevant to more than one component in the assembly. A start part was created with the root geometry modeled as datum features. The external geometry was created for the product, and then the product was divided into top and bottom housings. The top and bottom housing were used as root components for the flip-up door and battery door (not modeled). The modeling process paid particular attention to the assembly needs. In other words, the design intent was decided on at the outset of the project and was carried out in the design of the components. Attention was paid to ensuring that a quality fit between the components was achieved.

There are two other techniques for creating a product design of this type. The sections that follow explore how the other two approaches would aid in creating a product design, including the limitations of each approach.

Master Part Approach to Modeling the CD player

Assume for this example that you have modeled all of the geometry up to the point where the outside shape of the model is completely defined. Included in this geometry are the internal reference features described up to the previous phase 3. The external geometry is shown in the following illustration.

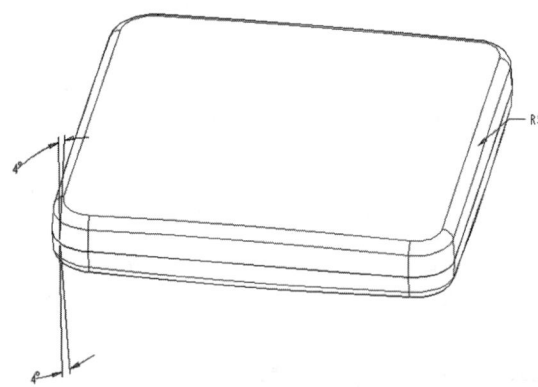

External geometry complete.

Design Philosophy and Master Part Plan

To begin this exercise, the master part will have within its structure all of the geometry that crosses more than one component. The idea behind this technique is that when the components are modeled they are actually cut out of the master part and referenced to its geometry. Using this technique, if changes were required to the external shape of the part, the designer would need to make those changes on the master part only. All of the components referencing the master would update upon regeneration. The following section explores what geometry could be used in the master model part plan for this project.

There are two things to consider in this type of modeling: what you want and do not want in the master part. Once the method is selected, the other issue to consider is an approach to this project that will take advantage of the master part's powerful attributes.

Looking at what geometry should be in the master model, you can approach the task by considering what geometry crosses more than one piece part or has to be referenced by more than one component. You must consider at this time that once the components are cut out of the master, further modeling is required to complete the components. Listing the features of the master model will help you apply the concept.

Desirable Master Part Features

You are creating a product design consisting of four external parts that reference internal geometry. The features it would be beneficial to carry in the master part are those that when once modeled may be used in more than one component. These features are much like the intelligent start part used in the previous exercise. The following are the features to note.

- The default generic start part with the default datums and standard accuracy
- The parting surface or plane
- Datum curve outlines for the keys and jack locations
- A surface structure to represent a space claim for the internal chassis, including the positions of the hold-down ribs
- The batteries as a surface model to provide a space claim for their position
- The datum curve profiles of the flip-up door and battery door
- All external geometry that will not fail the shelling of components or cross more than one component

Undesirable Master Part Features

Typically there are items in a design that cross more than one part and are therefore entities you do not want in your master part. The following items, which do not all appear in this modeling exercise, give you an idea of what not to model when creating a master file.

- Small external radii that would be a smaller value than the component wall thickness, causing the shell to fail.
- External styling grooves that could cause the shell to fail.
- Geometry used only on an individual component. This geometry does not cross more than one component.

Armed with the information of what you do and do not want in your master part, you can begin to model. The following section provides the modeling steps.

Model the Components

The master part as shown in the previous illustration is used as the first significant feature in each of the components. The components are cut from the master part merged with each component part file. A simple technique for this is to start a generic part and use the assembly tools. Assemble the master onto the component referencing the master geometry in an assembly. Call up the new component and cut the geometry for it out of the master file. The following are steps for creating the top housing.

1. Start a part using your generic start part, with default planes and coordinate system. Name this component *top_housing.prt*.

2. Create an assembly called *anything.asm* and add the new part to it.

3. Add the master part to the assembly.

4. Using the Advanced Utilities pick in the ASSEMBLY menu, select Merge. When prompted to select the part for performing the merge, select *top_housing* and then select Done. When asked to select the component to be used in the merge, select the master part. A simple way to ensure you select the correct part is to use the menus or model tree and select the component by name.

   NOTE: *To avoid duplication, do not copy the datum planes into the component when merging.*

5. Select Yes to remove the master component from the assembly.

6. Quit the window and call the *top_housing* component to the screen.

7. Cut away the geometry from the master to create the top housing part file.

This process may be repeated for the rest of the components, or you may simply copy this part to create the other components. The internal geometry may be added to each component, as in the previous modeling exercise.

Conclusions: Master Part Technique

Using the master part technique allows you to work with a single, common file for controlling the geometry shared by the components in the design. Designers need to be aware that geometry in this master part will be referenced by each of the piece parts. Any changes to this external geometry will be reflected in the component part once a regeneration of the component is initiated. The danger with this technique is that the geometry is dependent but the piece parts are not stand-alone parts. The master needs to be present in the directory containing the components to allow the components to regenerate properly.

Skeleton Model Technique

The skeleton model technique allows for shared geometry to be accessed in assembly mode. The skeleton model allows you to define geometry such as space claim geometry, datum curves, and location points used during the design process. In assembly mode, geometry may be copied from the skeleton model into individual components that are members of the assembly model tree. Geometry associated with the skeleton model does not get included in the mass property calculation for the assembly.

A skeleton model may be created in assembly mode using the design manager. If other components exist in the assembly, the skeleton will be placed at the top of the model tree. You may not add a component to an assembly and specify it as the skeleton model for the design.

To avoid repetition, the skeleton model for this design project would contain all features used as reference in the intelligent start part. When components are assembled, geometry can be passed from the skeleton model to the individual piece part files by means of the Copy Geom menu pick. This allows the designer to copy into component files only the geometry affecting a particular component.

A combination of a skeleton model and a master part with external geometry could be used only in an assembly to complete this design project. The basic geometry for each component could be cut from the master model using reference geometry from the skeleton model. The following illustration shows the features that would be included in the skeleton model for this project. The skeleton model would be placed at the top of the model tree structure in the assembly by the software, and features from the skeleton would be copied to the individual component files.

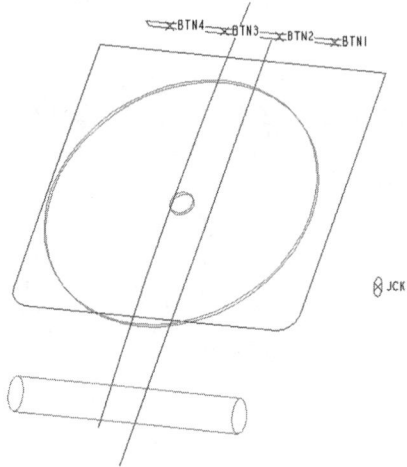

Skeleton reference model.

Summary

This chapter has discussed techniques used in a project design in which components need to share information. The techniques covered in this chapter were the use of an intelligent start part, master part, and skeleton model. One of these techniques may not be considered better than another. When planning the design task, consider each of the techniques and how they may be applied to your task.

Chapter 9

Replacing Several Components with One Plastic Part

Introduction

There are often opportunities when redesigning products to take advantage of new technologies. Plastic materials as a broad subject can represent quite a family of applications. Plastic is now even tough enough to be used to absorb impact in automotive applications. The array of materials allows the designer to replace metal parts, and sometimes even metal assemblies, with a smaller number of plastic components. The selection of material is most important when considering replacement of metal components with plastic. However, there are other factors to be considered, such as what role the metal part played in the design intent of the original product. You may not want to sacrifice design intent when trying to cut down on the number of components in an assembly.

To ensure that the properties of the material chosen will fulfill the design intent of an assembly, you need to consider whether or not a change to plastic will affect any of the performance standards for the product. For example, does the product need to meet flame test requirements? When situations of retrofit into current designs arise, it may be advantageous to consult with the plastic supplier to aid in identifying the opportunity and caution areas.

Project Description

In the project to follow, a metal frame assembly acts as the base for an electronic enclosure. The functions it serves are as a bottom on the unit and as a point of assembly for the side panels. This subassembly holds other components of the final assembly in place. The entire subassembly is attached as a unit to the outside housing to complete the assembly process.

In attempting to replace metal assemblies with plastics, you may find that a direct substitution of plastic for metal is not feasible. It is more important to look at the function of the metal assembly to determine if it is possible to provide a solution to the design problem without losing any of the attributes of the design intent. It is usually easier to apply a material change when you are given the freedom to redesign the entire product. The reason is that plastic components can provide a number of functions. With plastic you are able to use snap fit levers, seal-off ribs, ribs to guarantee space clearance between components, screw bosses, and surfaces that are cosmetically acceptable as exterior surfaces.

The assembly chosen for this project is a series of three metal parts and a top enclosure. The bottom and side panels serve only to enclose the electronic packaging. There is an existing top enclosure designed for plastic that serves as the overall cosmetic cover for the unit. The top enclosure may be modified if required to receive the new base component. The electronics are currently held in place with slides along each side panel. When the top cover is assembled, the electronics are not allowed to move. Therefore, the components are trapped on all four sides.

The following illustration shows the assembly to be replaced, with its metal base and side panels. The proposal in this situation is to replace the three metal components with one in plastic without sacrificing the design intent of the original sheet metal design.

Current sheet metal assembly.

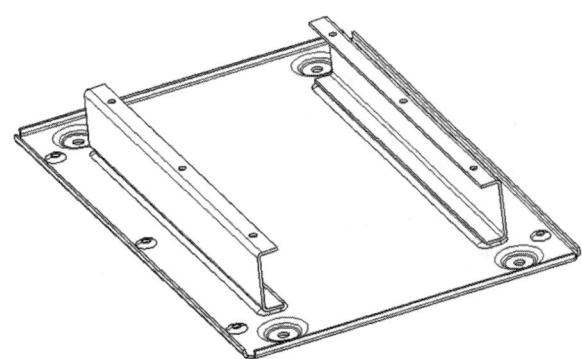

Design Planning

The sheet metal assembly shown in the previous illustration performs a variety of tasks. In replacing the assembly with a single component, you need to make sure you can duplicate or replace the features performing the various tasks, so that design intent is not sacrificed. This project is not as complex as, say, a bicycle wheel for which there would be a number of critical strength, roundness, and dimensional qualities to assess. However, you still need to identify what the assembly does to be sure you can meet the design criteria of the sheet metal design. Making a list of the assembly's attributes, such as the following, may help to identify how to model the component in plastic.

- The bottom plate acts to seal the unit from the bottom side.
- The side plates are used to mount the electronics board.
- The bottom plate also acts as one of the housings and joins to the top housing to close the unit.

With the previous criteria identified, you would be able to determine whether or not the same design intent could be met in plastic. The following list represents such an assessment.

- The base can still act as the seal for the assembly from the bottom side.

- The electronics board is still held in place provided it can be trapped between the top and bottom housings. This may require that other components be modified to allow for design rework.
- The joining of the housings can be accommodated by changing the mounting screw bosses.

You can conclude that if this design is to proceed you must also make some changes to the top cover of the unit. The major part of the redesign would be to attempt to complete a new lower housing that seals off the unit, holds the cover, and serves as the holding mechanism for the electronics. Fasteners would not be required to secure the circuit board in the plastic housings.

Modeling Plan

When creating a modeling plan for this redesign, you must also take into consideration the opposite housing, which must be mated to and sealed off against. The assembly to be replaced can be duplicated as a single part. You would first identify where the critical items are and model their locations. The electronics assembly must also be modeled (at least in primitive form), showing the mount locations and overall height restrictions. The following illustration of a primitive model for the electronics assembly shows the mount locations and height restrictions.

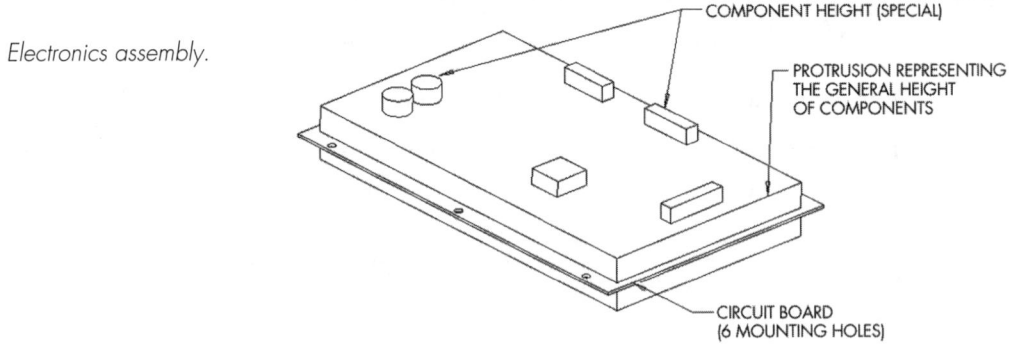

Electronics assembly.

The modeling plan would consist of identifying the holding positions for the electronics and proceed with the design from there. If you were able to place screw bosses at the circuit board mounts, you would be able to hold the electronics firmly in position using the base. Another option would be to provide hold-down ribs from the top housing and simply locate the electronics on the base housing. This would require that ribs be added to the existing top housing. If you were able to leave the top housing mount locations where they are, you would not need to move any screw locations in the top housing. Stiffeners might be added to the plastics to help strengthen the component.

Modeling Process

The following are the phases and steps involved in modeling the electronics housing.

EXERCISE

Phase 1: Create New Component Containing Datums

The steps for phase 1 follow.

1. Create a new component using the generic start part to obtain the default datums, view names, and accuracy setting. Add a datum plane to this part to represent the mounting plane for the electronics.

2. Create six datum points to represent the hold-down positions of the electronics. On the bottom datum plane, create six datum points to locate the housing mounting positions. On the electronics datum plane (labeled BOARD in the following illustration), create a datum curve to represent the exterior of the electronics. With these items located, the design of the actual plastic part may proceed.

   ✓ **TIP:** *This component may be assembled onto the electronics component and top housing to aid in designing the components. The assembly would also serve as the mechanism from which the dimensional values of the piece part may be derived.*

The following illustration shows the datum features added to the start part. Notice that the datums are labeled for purposes of identification.

Start part with datums added.

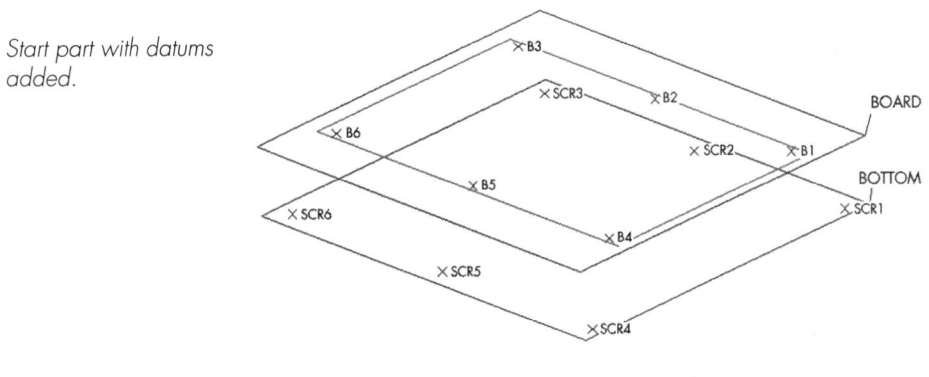

Phase 2: Create the Base Design

The base design must be created such that it mounts to the top housing in the original screw locations. The base feature will have the same external profile as the sheet metal component. The screw locations will be recessed to take advantage of the current locations of the top housing bosses. The steps for phase 2 follow.

1. Using the bottom datum plane as reference, create a protrusion with a thickness of 10 mm. The external dimensions of the protrusion may be taken from the sheet metal component to ensure the appropriate clearance in the assembly.

2. Create draft from the top surface of the base using the top surface itself as the reference plane for the draft feature. Dimension the draft such that the outside walls of the component become smaller as the draft moves away from the reference surface. The draft value is 4 degrees.

Design Planning

✓ **TIP:** *Use the Boundary Loop option to select all of the external surfaces at once, and select the top surface as the reference surface. This will allow all of the surfaces bounding the reference surface to be selected in one menu selection.*

3. Create the protrusions from the bottom surface, which will be used for the feet mounting. The four rectangular pads sit 20 mm from the external walls and measure 15 mm by 15 mm. The protrusion height is 4 mm from the bottom surface. The following illustration shows the protrusion, draft, and footpad added.

Protrusion, draft, and footpad added.

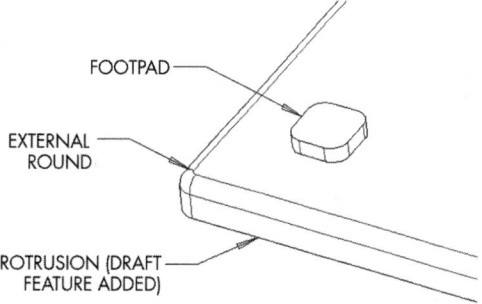

4. Create rounds on the rectangular pad to remove the sharp edges on the four corners of the rectangle. The value of the round is 4 mm.

   ↪ **NOTE:** *The geometry being constructed is geometry you want included in the calculation of the shell feature. If you were to model a radius value of 1 mm on an outside edge, it would result in a shell failing if the shell value was greater than the round value.*

5. Add draft of 4 degrees to the external walls of the rectangular pads.

6. Add corner radii of value 3.5 mm to the footpads on all of the remaining edges.

7. Add external rounds of value 3.5 mm to the top and bottom edges of the footpad. The completed footpad is shown in the following illustration.

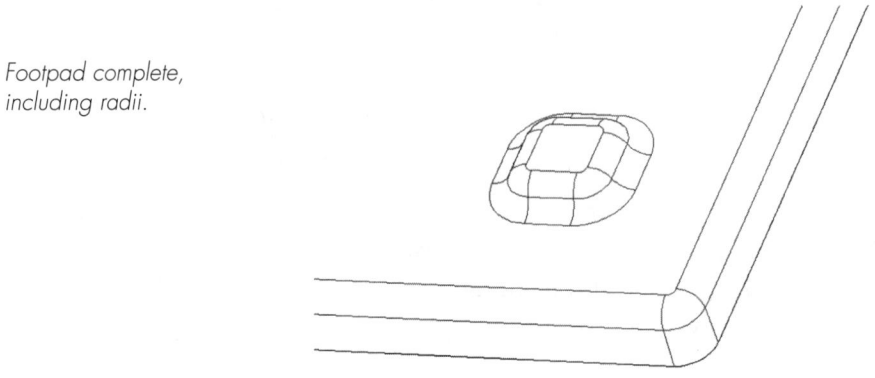

Footpad complete, including radii.

Phase 3: *Pattern the Footpad and Create Duplicates*

The footpad is now complete in terms of its external geometry. The features of the footpad modeled as individual geometry can now be combined and patterned to obtain the other three footpads. The important thing to note is that all of the external geometry that will help prevent a shell failure has been created prior to modeling the shell feature. If you were to proceed with a shell, leaving a wall thickness of 3 mm, it should not present a problem, as none of the external geometry is less than the thickness of the shell.

The features that constitute the footpad have been modeled sequentially. A group may be made of the sequential features and patterned to obtain the total of four pads. It is your choice whether to simply pattern the footpad or to create a local group first and pattern that. If you choose to employ the Pattern option, the pattern will be difficult to change later in the modeling process should you want one of the patterned features to be different than the rest. By first making a local group, you are able to unpattern the patterned features and modify then independently. Any dimensions chosen as patterning dimensions may be made independent by unpatterning the set of four footpads. The steps for phase 3 follow.

Design Planning

1. Create a local group with the footpad protrusion, draft, and external rounds included.

2. Select the vertical placement dimension as the pattern dimension in the first direction, with an increment distance of 145 mm. Select the horizontal placement dimension as the pattern dimension in the second direction, with an increment value of 245 mm. Both directions will have the number of instances set to a value of 2.

3. Select Done and the pattern of four footpads will be completed, as shown in the following illustration.

Footpads patterned.

Phase 4: Shell the Geometry

With the footpads in place, you can proceed to shell the geometry to a constant wall thickness of 3 mm. The surface to be removed by the shelling process is the top surface. With the shell in place, you can proceed with the rest of the internal modeling requirements. The step for phase 4 follows. The illustration that folows the step shows the completed shell.

1. Create a shell feature with a wall thickness of 3 mm. Select the top surface as the surface to be removed during shelling.

Shell complete.

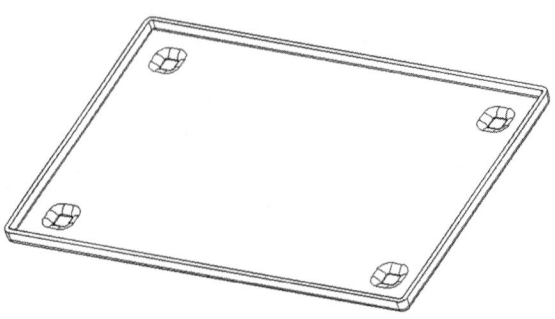

Phase 5: Model the Screw Boss Mounting

The mounting screw locations are predefined on the existing top housing, and you have established datum points at each of the boss locations. The elevation of the mating surface of the screw boss faces is 14 mm from the bottom datum plane. To allow the screw boss faces to be the last surfaces pulled together when the screws are put in, you will model the screw boss height at 13.8 mm. This dimension will allow the assembly to be drawn securely together when the screws are tightened. The steps for phase 5 follow.

1. Create a protrusion using a make datum plane at a value of 13.8 mm off the bottom datum plane. The protrusion has a diameter of 9 mm.

2. Apply draft of 1 degree to the external walls of the protrusion.

3. Create a revolved cut on the center line of the protrusion for the screw, as shown in the following illustration. The cut will be revolved 360 degrees.

Design Planning

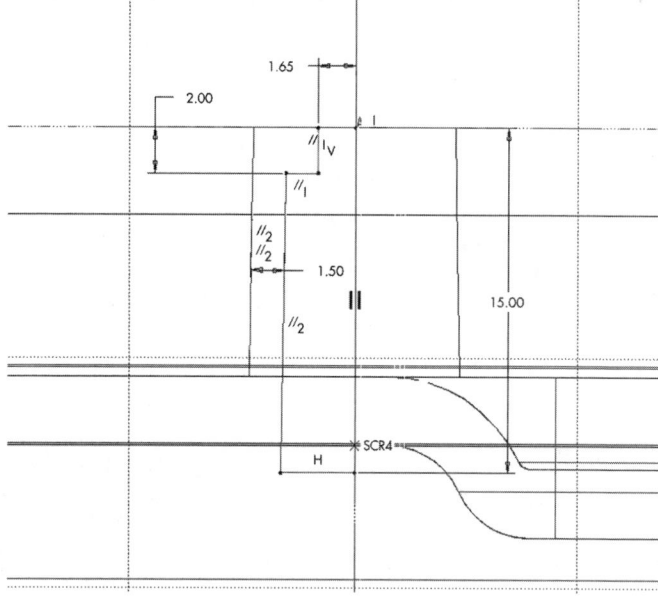

Sketch of revolved cut for screw.

4. Add finishing rounds to the base of the boss for strength and to aid in material flow.

5. Pattern the screw boss features or copy them to the other five locations using the same technique used to pattern the footpads.

Phase 6: Place the Electronics Mounts

The mounts for the electronics are taken at the height of the electronics plane. The mounts will consist of a series of ribs around the perimeter of the electronics, as well as pins in the area of the mounting holes. Hold-down ribs will be added to the top housing to complete the trapping of the electronics assembly. The steps for phase 6 follow.

1. Create a protrusion of 6-mm diameter posts in the six locations representing the electronics board mounts.

2. Create pins of 3.5-mm diameter on the top of the posts to align with the holes in the electronics package. The length of the pins may be 2 mm. A finishing round of 0.5 mm may be placed on the top of the pin.

3. Create a cut of 2.5-mm diameter to core out the bosses. Use an offset datum plane 2 mm from the electronics plane.

4. Add 1-degree draft to the internal and external surfaces of the six hold-down bosses. The hold-down posts are shown in the following illustration.

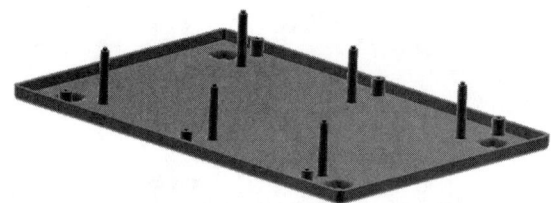

Hold-down posts added.

5. Add ribs to control the side-to-side movement of the electronics package. The ribs should be 1.2 mm in width at the height of the electronics plane and have draft added to them at a value of 0.5 degrees. The ribs may be modeled in three positions along the side of the electronics package and in two positions along the ends. The cross section shown in the following illustration indicates the shape of the finished ribs. Creating one rib and patterning it to the other locations would best create the ribs in both directions across the part.

Side rib cross-sectional view.

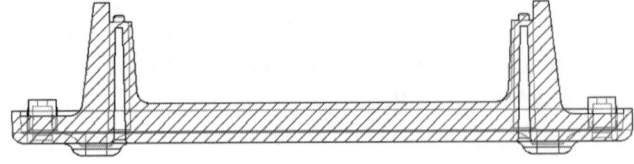

These ribs are common in form in each direction and may be patterned once one rib is created in each direction. Once the ribs are in place, radii at a value of 0.5 mm are required along the edges where the ribs intersect the main housing. Hold-down ribs would also be required on the top housing to clamp the electronics assembly in place. The finished lower housing is shown in the following illustration.

Finished lower housing.

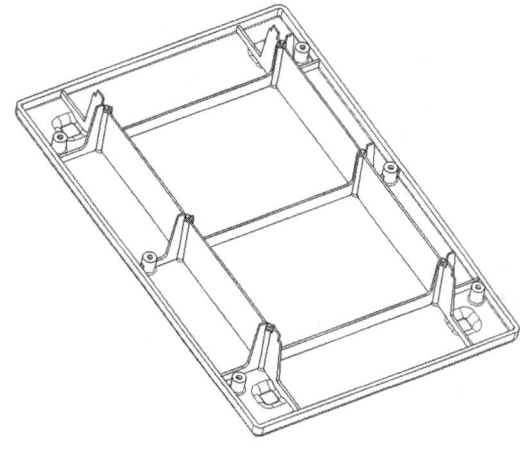

Summary

Pro/ENGINEER provides the designer with tools for creating complicated parts that replace a number of individual parts. By managing reference locations and dimensioning to fixed datums, the designer can redesign products with confidence that the original design intent of the product is maintained.

The approach employed in the exercise in this chapter replaced three sheet metal parts with a single plastic housing. Standards for the electronics enclosure, as well as specific material properties of the material chosen, would need to be considered to ensure success of the project. In this exercise, the top housing would have required some modification to ensure that the product was securely fastened together in the final assembly process.

Chapter 10

Component Fastening

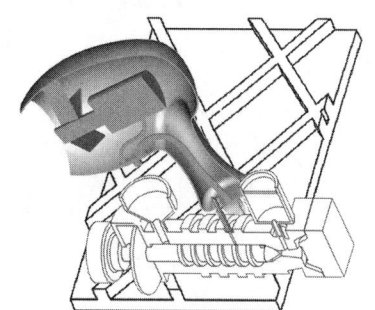

Introduction

Plastics technology offers a number of methods the designer can employ for fastening components together. Pro/ENGINEER, because of its solid modeling capabilities, enables you to explore component fastening in a realistic manner. Pro/ENGINEER's tools for modeling and viewing assembled components are among the best in the world. The software even allows for realistic interference checking between components to ensure that your designs are meeting expectations.

This chapter focuses on the technology of putting parts together. You will explore a number of techniques for assembling plastic parts and how Pro/ENGINEER can aid you in modeling design requirements in the most appropriate manner. The technologies covered aid in installing piece parts in a product, as well as fastening the overall product assembly.

Holding Components in Desired Locations

Components should be designed such that they do not rattle when installed in the intended position. The quality of the overall product design may be excellent, but if the internal components are allowed to move around even a little, the consumer may decide that the product is of inferior quality. The previous two chapters and Chapter 4 have explored pinching or clamping components in position while installing screws. This works well for the "Z" direction of an assembly, but what about controlling the side-to-side movement of parts?

If you had a feature or set of features that held internal components firmly, resisting side-to-side movement, the overall quality of the design may be perceived to be better than one that did indeed rattle or shift a little. This design problem can exist in many designs with internal components, and is amplified if internal components are heavy.

The following sections consider a couple of techniques for holding components in place. One technique is heat staking components in position by softening the plastic and forming it to hold the internal piece in place. Another technique is to use clamping in the "Z" direction, with crush ribs used to prevent the component from moving laterally.

Heat Staking

You can fasten an internal component to remain in a constant position in a variety of ways. You can glue it in place (which may not be environmentally friendly), you can use screws to hold it in place, or you can place tabs or pins on the mating component and simply heat stake them to secure the component. In the third case, you would have to break the joint and destroy the fastening mechanism to remove a component. This needs to be taken into consideration if the unit you are designing may require repairs in the field, thus requiring disassembly.

Heat Staking

Staking Pins

When using pins to provide the securing mechanism, you need to consider the number of pins necessary. The result of the process is roughly the same as if you were using screws. Pressure will be placed on the assembly, the pins heated and riveted, and a cooling cycle gone through before the assembly is finished.

In Pro/ENGINEER, you would obviously model the pins in their original form prior to staking. This technique holds components firmly in all directions within the capacity of staked pins to resist the forces placed on them. Heat staking pins are shown in the following illustration.

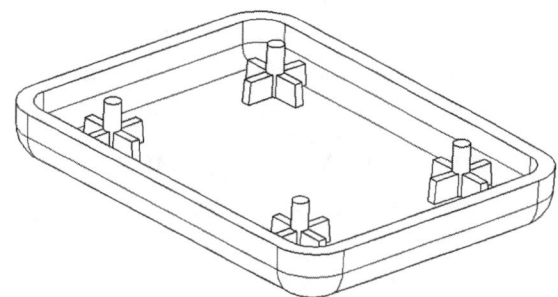

Heat staking pins.

Everything has a breaking point. To prevent pins from shearing off if an excessive side load is exerted on the assembly, consider surrounding the component you are trying to secure with ribs to help take load from the side direction. In addition, perhaps a combination of ribs and clamping is indicated. Ribs may be used in conjunction with clamping a component in the "Z" direction. In fact, you could purposely design the side load ribs such that they always apply some holding force to the assembly. These are called crush ribs. Note that heat staking does not apply to just round pins but to other types of heat staking features, such as ribs and tabs.

Crush Ribs

Crush ribs seldom become a physical reality in a component or part the way they are designed. When employing crush ribs, consider steel safe modeling and fine tuning of the fit after you have received and checked out the first parts. The technique for using crush ribs is relatively straightforward. You intentionally design interference fit between the internal components and the positioning mechanism that holds them in place. When the assembly is drawn together, the components force their way into position by crushing or slightly shearing off the plastic in the path of their travel. This is another situation where you would purposely design interference between components in the finished file. You may want to keep the ribs just short of where you think they should be until you examine your first plastic parts.

> **NOTE:** *It is always a good idea to involve the toolmaker when you intend to model features that require fine tuning after the first parts are examined. Working with the tooling supplier to identify areas that might change can save a considerable amount of rework cost associated with a mold. Areas identified as highly likely to change may be inserted by the molder to allow for easier change.*

Crush ribs should be modeled such that they allow for the alignment of the internal component prior to the actual crushing of the ribs. You would not want to force something together that was not in its desired position. A simple lead-in or chamfer will suffice to allow the component to partially seat in position. Crush ribs should also follow the 60-percent rule for thickness. Having crush ribs so huge that they give you sink marks on the external surfaces of the plastic can result in products with a poor appearance.

Rather than thick ribs, consider using more of them. Pro/ENGINEER patterning tools can quickly duplicate the rib design in more locations. Another trick to simulate a thicker rib would be to have a pair of ribs in the desired location rather than just one. This will give the effect of a thicker rib without giving you the sink mark at the rib location. The following illustration shows the characteristics of a crush rib with intentional interference. Keep in mind that when checking for part interference in your assembly you will find it between the two components.

Crush rib design.

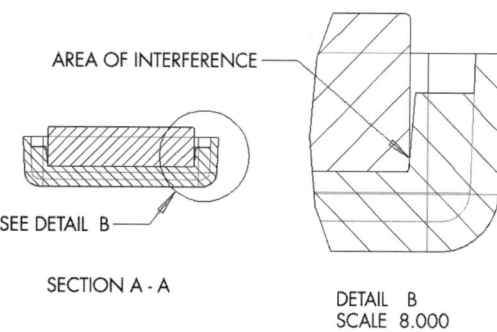

It is important to note that the process does not shear 1, 2, or 3 millimeters off the ribs. A minimum amount of interference material (0.1 to 0.3 mm) is used in the crushing process. As the previous illustration indicates, to ensure the correct seating of the component, you must also allow somewhere for the material to go if the component shears it off.

Ultrasonic Welding

Ultrasonic welding of components may be desirable if your design is intended to be tamperproof or sealed in some fashion. What you are actually doing when welding components is melting a small plastic rib and allowing the molten plastic to form a bond between the components. There are a number of techniques to use for this type of fastening. This discussion focuses on ultrasonic welding of the perimeter of assembled components.

It should be understood that the welding process might also affect internal components. If there are internal plastic components intended to have motion, or slide, there is a strong possibility that they also could weld during the welding cycle. It is advisable to consider consulting an expert in the field of ultrasonic welding when designing piece parts and planning the intended tooling and manufacturing process.

Understanding the Process

Prior to looking at the design of the welded joint in Pro/ENGINEER, you need to understand what it is you are trying to accomplish and to take a quick look at some common pitfalls. Take an example of a top and bottom housing with a subassembly placed within them. The goal of the welding process is to complete a bond between the two external housings while trapping the internal subassembly. To accomplish this, you need to identify how the welding will take place. To simplify the example, assume that the bottom housing is placed in some sort of assembly jig (nest). The internal subassembly is placed into the bottom housing. The top housing is set in place on top of the unit. The following are design considerations.

- The assembly jig is designed to be the holder for the bottom part of the unit during the welding operation.
- The internal components are placed in position.
- The top housing has been placed on the assembly.

At the welding station, a horn will be positioned on top of the top housing. The horn has been designed to mate with the surface contour of the top housing, particularly in the area of the intended welded joint. From this point, a load is placed on the assembly, and an ultrasonic vibration induced for a specified period of time in the horn while the horn is lowered a predetermined amount. The assembly is then held in position to cool after the horn has been turned off and before full pressure is reduced. The weld is then complete. The following illustration shows a welding jig layout for the video game controller.

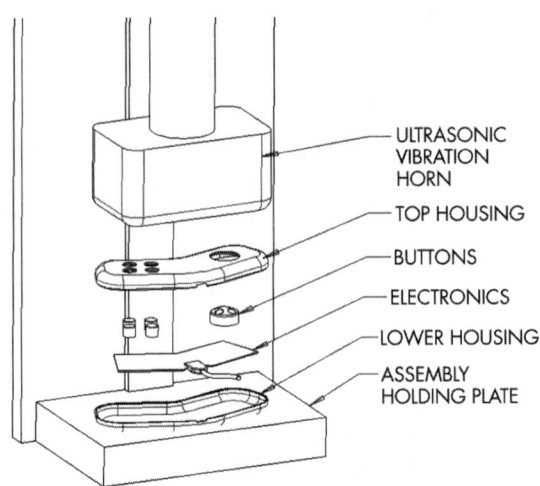

Ultrasonic welding jig layout for video game controller.

Any areas of interference in the plastic parts will also tend to be welded (melted together). Because of this, ultrasonic welding may not be appropriate for a design in which the following situations might hold true.

- Ribs intended to apply internal pressure to another component may have their ends melted off, resulting in relaxation of the intended pressure.
- Sliding fits may weld permanently.
- Snap fit joints may weld.
- Heat staked joints may relax if under pressure.
- The assembly is not easily taken apart. Disassembly results in destruction of the housings.

Keep in mind that some types of plastic do not lend themselves to ultrasonic welding as well as others. Your plastic supplier can help with the material selection and with the design of the application.

Butt Weld Joint Method

Good joint design is important to the success of a welded assembly. Pro/ENGINEER provides tools that allow you to create the features necessary to a successful welding task. The following illustration shows a cross-sectional view of a joint without welding features installed. This view indicates how you want the finished product to be seated. The top and bottom housings will have an interlock rib for alignment and rigidity of the assembly. The lip feature has been used to create the interlock mechanism. The components will in fact seat against each other during the assembly process.

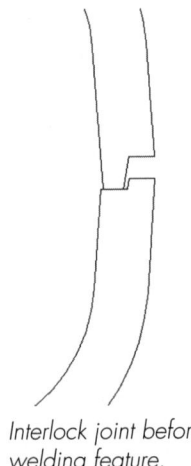

Interlock joint before welding feature.

A butt joint will add material that will be melted during the welding process, resulting in the finished joint shown in the illustration. Therefore, the material will be intentionally added as interference between the parts. The welding process will melt the material and the stroke of the horn will displace the top cover until it meets with the mating parts. This is typically not a huge amount (0.15 to 0.3 mm) of displacement. The feature may be added as a swept feature around the perimeter of the component on the interior of the part. For the purpose of cosmetic appearance, this will keep any melted material inside the component assembly.

On the bottom cover, create an advanced variable section swept feature along the internal trajectory of the bottom cover. The secondary trajectory is not required when using a pivot direction plane to ensure the feature is manufacturable. To save modeling steps, include the draft in your sketch for the protrusion, and bury it slightly in the solid if you are sweeping along a curved profile. This will help the software calculate the intersection. The section for the swept protrusion would look something like that shown in the first of the following illustrations. An alternative technique would be to use the lip feature and offset the surface. The second of the following illustrations shows the finished butt joint, indicating the area of interference.

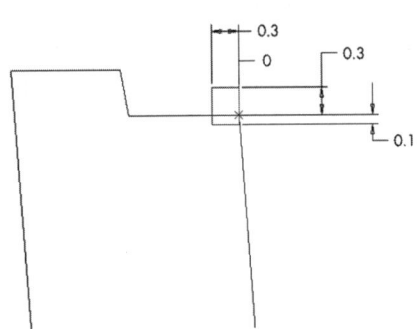

Section for the advanced variable section swept feature.

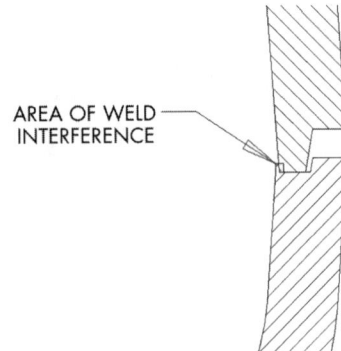

Finished butt joint feature showing area of interference.

Ultrasonic Welding

The welding process will remove the interference material at the joint between the housings. Therefore, for modeling purposes, your assembly will have interference. It is typical during design to leave this feature until the design is nearing completion. The following illustration shows the joint in place. The overlapping material, which will be melted to form a bond, must be adequate to hold the assembly together. It is important to note that any molten material from the process that runs will probably go inside the assembly.

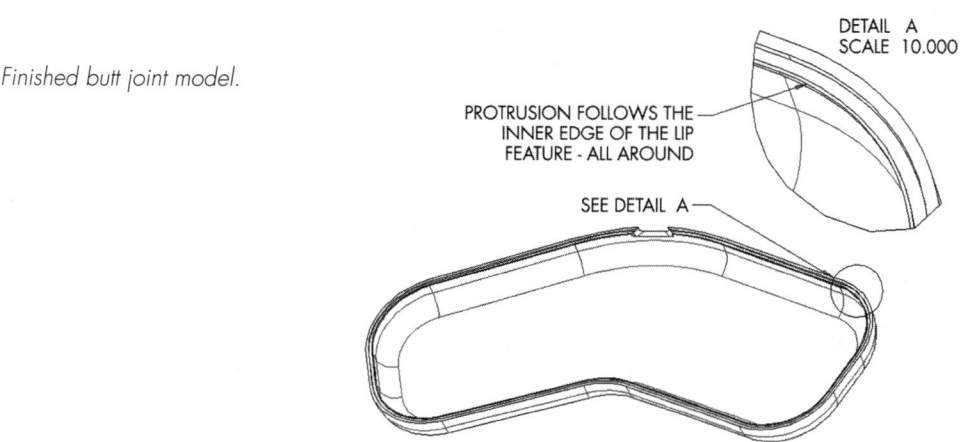

Finished butt joint model.

Energy Director Method

The energy director method follows basically the same principal as the butt joint. The process for modeling the feature in Pro/ENGINEER is similar to the butt joint, having a single trajectory and using the pivot direction plane to keep the finished feature manufacturable. The energy director will also concentrate the vibration from the assembly process in a specific area of the design assembly. The following illustration shows a sketch of the energy director swept protrusion. The feature will be slightly buried in the existing solid.

Sketch of the energy director swept protrusion.

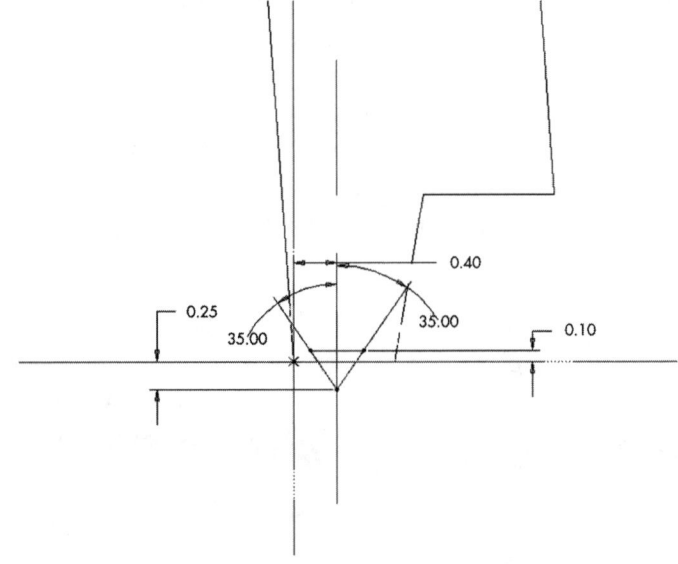

The modeling of this feature is very similar to the butt joint except that the feature itself follows the path of the original trajectory rather than lying directly on the path line. Consideration may have to be given in your design to where the melted material might flow to ensure that the melted material does not become visible from the external side of the assembly. In this case, a relief might be added to act as a trench for the melted material to flow into. The trench could be modeled into the mating part as a cut using the same principle as the protrusion. The resultant assembled joint would look as it does in the following illustration.

Energy director and relief trench.

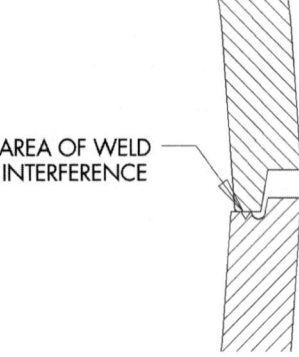

Snap Fit Design

The snap fit design requires some method of primary alignment of components to ensure the design intent is followed. A snap fit joint provides the fastening mechanism between components. It is a good idea when dealing with snap fits to model common features into mating parts for ease of manipulation of the finished features. Snap fit designs typically require that slides be used in the mold to form the actual snap feature. Snap features rely on the flexibility of the plastic material to make them work properly and would be best modeled using steel safe dimensions until the first parts are inspected.

In order for the joint to hold the assembly firmly without the components rattling, the joint should always be under some sort of load. Because of this, snaps are better located in less rigid parts of the housing to allow for over-travel during the assembly process. The following illustration shows a rectangular housing with examples of snap locations. In this example, snaps are located too close to the rigid corners at the top of the housing. The bottom of the housing would be a preferred location for snaps, which would allow the plastic housing to deflect during assembly.

Snap locations.

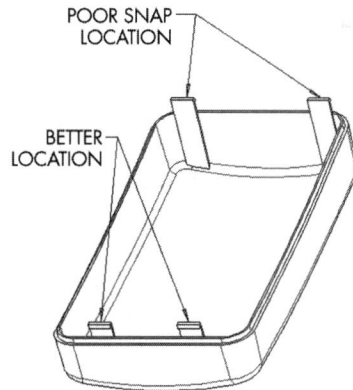

When modeling snap locations, you can use a datum point to locate the center of the snap feature. As a result, all of the related snap features will follow the point if the center of the snap needs to be relocated.

Modeling Process

The following are the phases involved in modeling the snap-fit housing.

EXERCISE

Phase 1: Locate the Snap on the Top Housing

Place a datum point along the external edge of the part at the desired snap location. An offset datum plane (18 mm) may also be modeled to help locate the point along the edge of the part. Repeat this process on the bottom housing. If the snap feature is moved, the datum plane offset dimension is the only thing that requires modification. The following illustration shows the located center point.

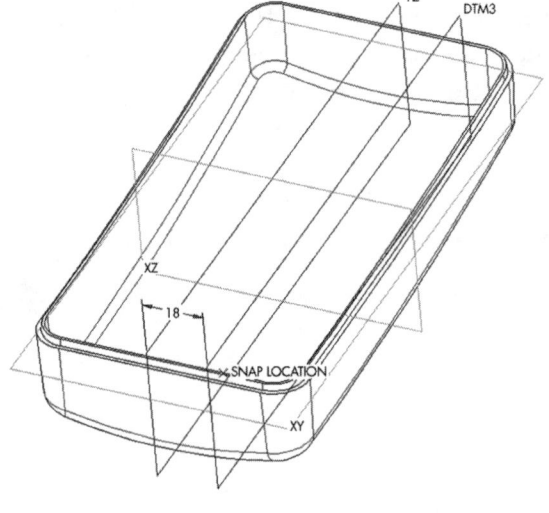

Center point located.

Phase 2: Add Snap Protrusion

Create a protrusion on both sides of the datum plane, modeled through the snap location point. The protrusion will reflect the male portion of the snap feature, shown in the following illustration. The draft may be added directly into the sketch for the protrusion or left as a separate feature. The lead-in angle for the snap may be modeled directly into this feature or left as a chamfer feature to be added later. For the purpose of this example, the lead-in chamfer is included and the draft is left until later. The protrusion (snap) width is 9 mm.

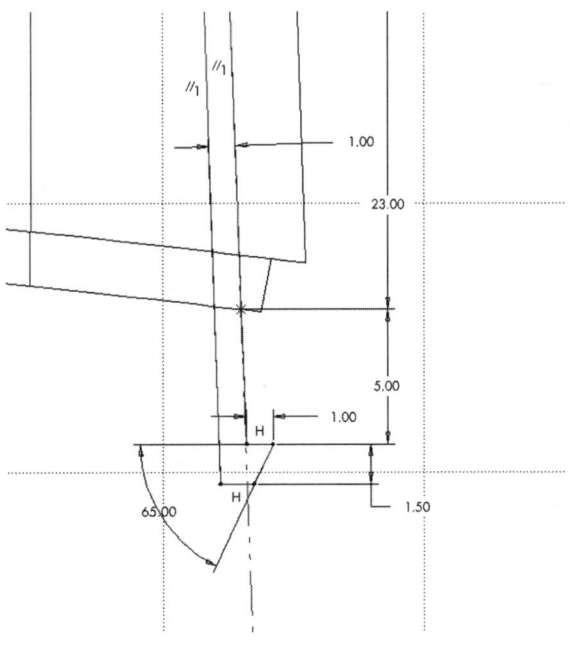

Sketch of male portion of snap protrusion.

Phase 3: Add Draft and Finishing Rounds

Draft and finishing rounds may be added to the snap protrusion to finish the feature design, as shown in the following illustration. The rounds placed at the base and sides of the feature will not only help the material flow but will add to the strength of the snap. The rounds may or may not be able to be patterned when copying this feature to another location. If this is the case, remove the rounds, pattern the feature, and add the round features later.

Draft and rounds added to the snap.

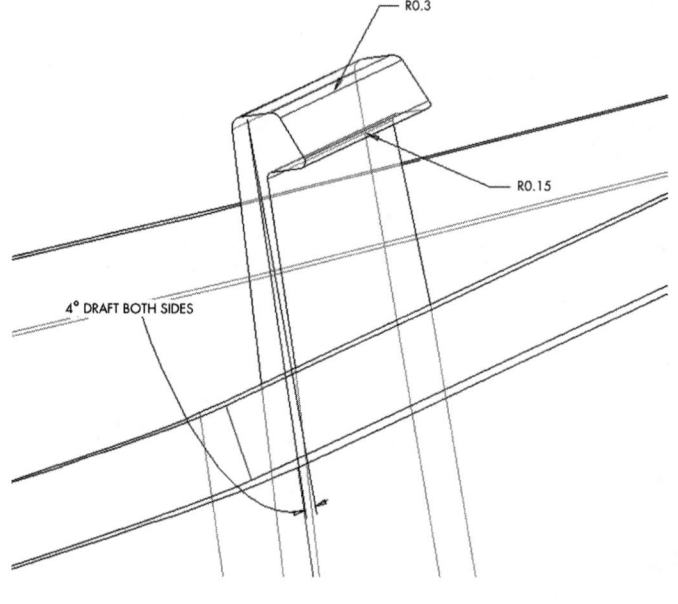

Phase 4: Pattern the Features

The features of this model, required to make the snap protrusion, have been created in sequential order. As a result, they may easily be patterned to form more than one location for the snap. Using the Group menu pick, create a local group consisting of all features that constitute the snap, starting with the offset datum plane. Pattern the group using the offset dimension for the datum plane as the increment dimension for the first direction. Ignore the second direction pattern dimension, which is not required in this case. The increment for the dimension in the first direction is defined as –36 mm. The number of copies would be two. The finished male snap features are shown in the following illustration.

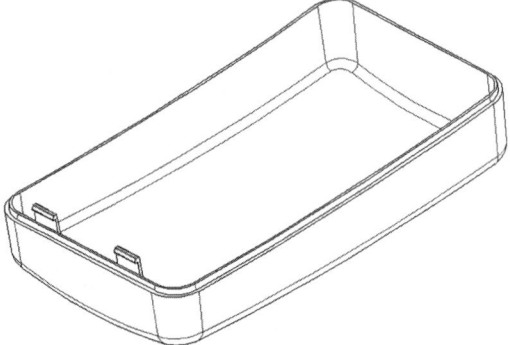

Finished male snap features.

Phase 5: Creating the Snap Hole

When choosing snap locations, you must take the surrounding geometry into consideration to ensure the snap design is manufacturable. If slides are required in the mold to produce the snap geometry, there must be room in the design for the slide to operate in the mold.

Create a cut on both sides of the datum plane, locating the center of the snap feature at a width of 16 mm. This will give ample clearance for the side-to-side tolerance of the snap. The following illustration shows a sketch of the cut.

Sketch for the snap cut.

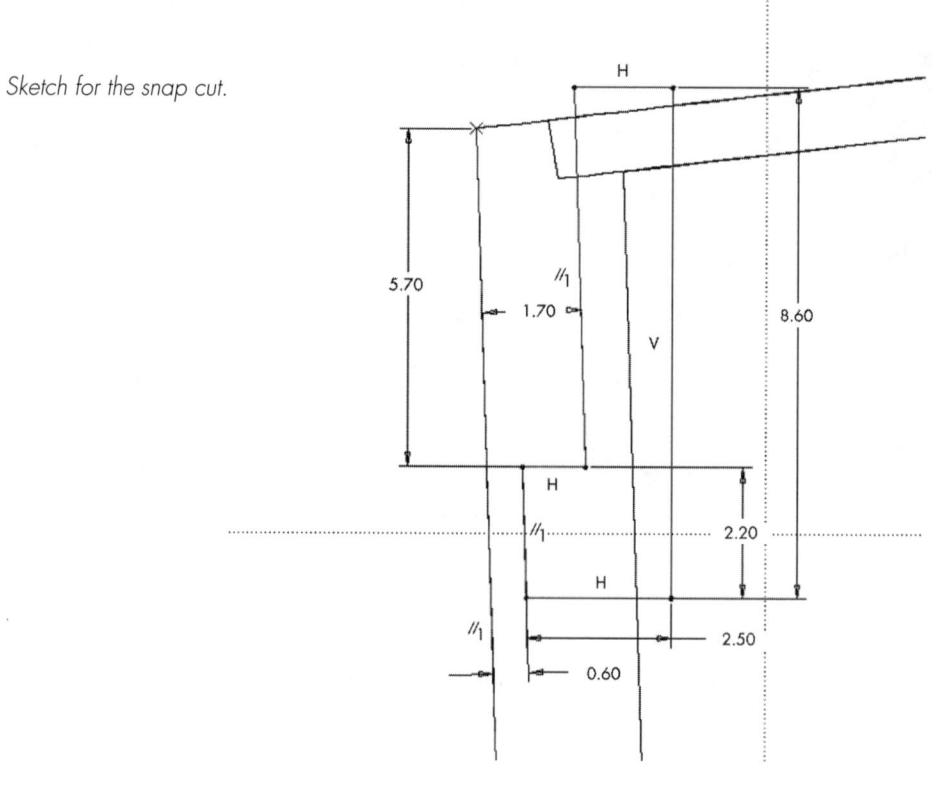

Phase 6: Finishing Features

The draft must be added to the internal walls of the hole created by the snap cut. The draft will be created referencing the line of pull of the slide in the mold. A 1-degree draft may be applied to the inner walls. A datum plane, shown in the following illustration, has been created to show an appropriate location for the draft plane. Finishing rounds may then be added to the inner corners of the hole. A lead-in chamfer for the snap may also be modeled if desired or required.

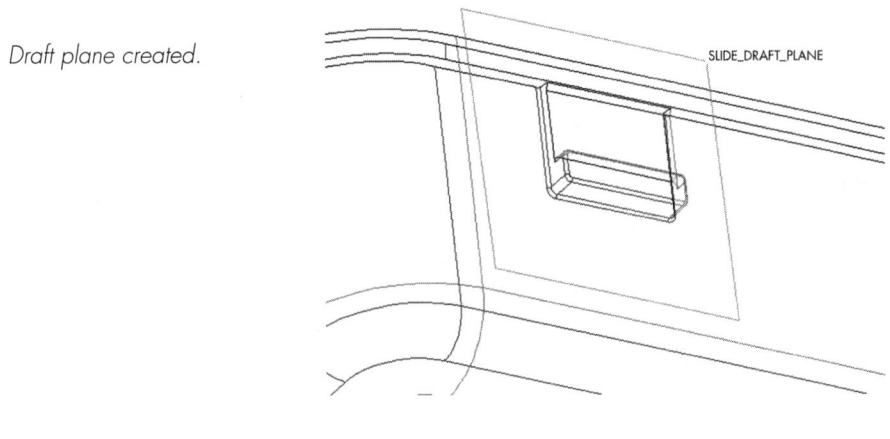

Draft plane created.

Finished Snap

With the female snap hole cut, you are able to pattern it as in the previous example to make the second copy. The following illustration shows a cross-sectional view through the housing assembly at the snap location. The view is marked to show the line of draw of the slide in the bottom housing and the face on the male snap that should be modeled steel safe and tuned up after the first parts are manufactured.

Cross section of finished snap design.

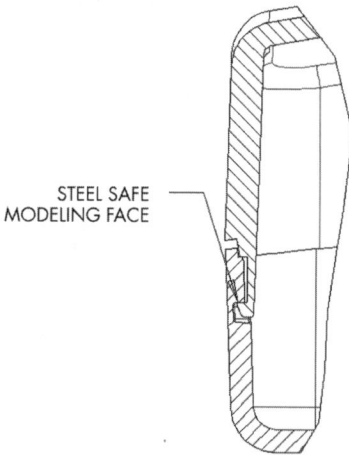

Screws

The project in Chapter 8 showed how to use screws as fasteners in product designs. Looking at some preferred practices will help identify opportunities to look for when using screws in a design. How many screws will hold the product together suitably and what size the screws should be can be a bit problematic to determine. For economic reasons, there may be standard or preferred screws your company purchases in bulk, or you may have standards particular to the tooling on the assembly line, which will force you to choose screws from the list available.

Whatever the screw used, the amount of holding force of each screw is relative to the plastic material used, the type of screw chosen, and the amount the screw is torqued down during assembly. If in doubt, work with the manufacturer's guidelines for plastics when determining the number of holding positions required and the screw size.

The following are points to keep in mind when assembling plastics modeled and managed in Pro/ENGINEER. Recall from Chapter 8 the assembly order for clamping components. As screws are tightened into seating position, there are items that need to come together in a specific order to ensure a robust assembly.

Snap Fit Design

- The housing interlock should come together first.
- Hold-down features for internal components are the second items to clamp together.
- The last thing to clamp together is the screw boss seat.

The following illustration shows the order of clamping as screws are tightened. The dimensions have been exaggerated in the illustration to better show the sequence. Datum planes have been modeled to control the clearance in the hold-downs and screw boss seats. If more clearance is desired, the only thing the designer would need to do is move the datum plane controlling the features and all of the related features would update automatically.

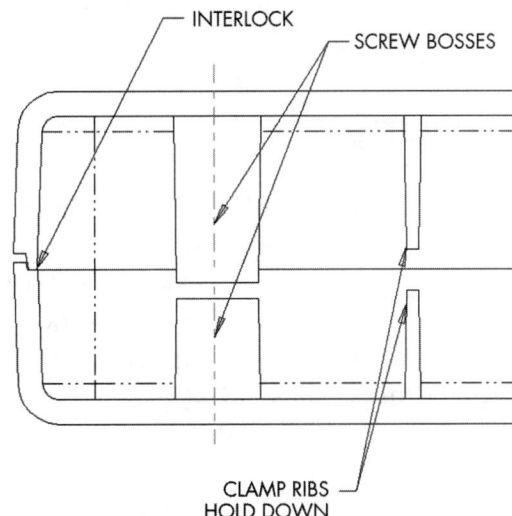

Sequence of plastic assembly clamping using screws.

When using screws for housings it is best to consider the location of screws early in the design process. Placing the screws on the external vertical walls can create difficulty in maintaining a suitable wall thickness of material, as well as hamper your ability to use the screws' clamping capabilities in the most effective manner. The best place for the screw is likely between the external wall and the hold-down features, as depicted by the left-hand image in the following illustration. The right-hand image in the illustration indicates that when the screw is placed on the wrong side of the hold-down ribs the ribs may act as a pivot and lift the interlock apart while the screw is being clamped.

Placement of screw bosses.

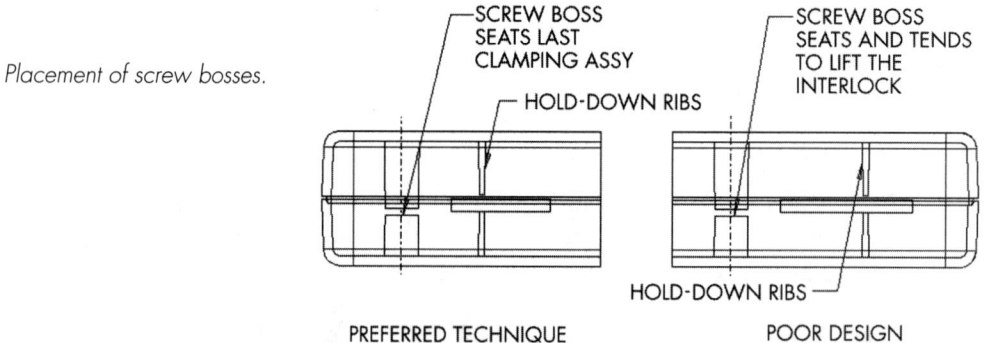

Summary

This chapter has dealt with various techniques of joining plastic components and assemblies. Heat staking, crush rib design, and ultrasonic welding are three techniques to be considered. Screw boss location and fundamental application rules were covered to convey sound product assembly practice. The chapter also covered the concept of snap fit features and how to manage this type of modeling in Pro/ENGINEER.

Chapter 11

Thin Wall Component Design

The Special Molding Technique

Introduction

This chapter covers some of the modeling techniques for creating a plastic housing when thin wall technology is chosen for a project. The rules on rib and web creation will be looked at and how they apply to thin wall design. Pro/ENGINEER analysis tools will be used to monitor the wall thickness of the design to ensure it is kept to the desired size. This chapter also introduces you to rib design techniques used to strengthen a thin wall part design. Consideration is given to how draft angles affect this type of design when modeling external and internal part geometry. Once the modeling is complete, the part is tested for integrity using some of Pro/ENGINEER's analysis tools.

What Is Thin Wall Design?

"The Special Molding Technique" has been used as a subtitle for this chapter to specify the technology referred to in this chapter. Thus the question, "What is thin wall design?" Does thin wall molding refer to simply making components thinner than you do now?

Conventional molding technology can quite often be used to create components with thinner wall thicknesses than their existing forms. For example, if the typical wall thickness used by your company is 3 mm and you want to make new components 2.5 mm, you should be able to accomplish this without using thin wall molding technology. Technology advancements such as sequential gating and computerized process control in conventional molding have improved with time and have advanced the technology to the point at which some component designs may not require thin wall technology to produce parts.

Thin wall technology requires special equipment with higher clamping forces in the molding equipment, as well as special sprue sizes to deliver the smaller amounts of plastic to the mold. Because the plastic injection pressure is higher in the mold, the material flows in much more quickly. Due to the thin wall of the components, it is essential to fill the mold in a shorter period (before the plastic freezes off). Therefore, the mold pressure is set higher. This results in some of the plastic design rules being relaxed because of the resulting pressure of the molding process. Rib thickness may be the same as the wall thickness in thin wall design, which greatly reduces the likelihood of sink marks. Draft, on the other hand, becomes quite critical and the designer should endeavor to stay on the high side of the draft angle where possible, using 2 degrees both internally and externally on the part.

Design Philosophy

The design chosen for the project in this chapter is a plastic housing for a paging unit. Very small in physical size, the product must hold an electronic board and viewing screen. Note that thin wall molding will allow a wall thickness of 0.8 mm for the bulk of the body of the housing and 1 mm for the sides. The exercise that follows employs a constant wall thickness of 1 mm. Alignment ribs will control the fit of the plastic housings and there will be one screw in the center of the unit. The screen is located at the top end of the unit. The battery is included with the circuit pack and not accessible to the customer.

Modeling Plan and Process

The following illustration shows the design features and characteristics of the pager assembly. The exercise that follows consists of modeling the pager's front housing.

Pager assembly.

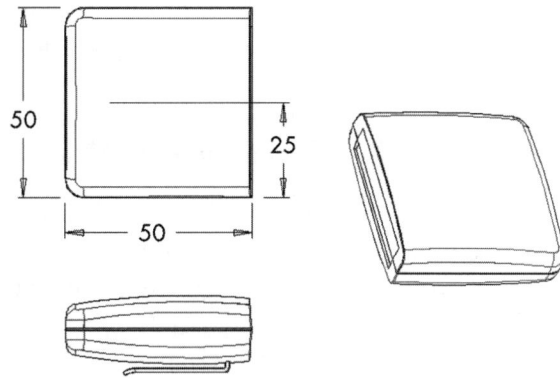

Modeling Plan and Process

Modeling Plan

The model of the front housing for the pager unit has a number of identifiable characteristics you might list to help evaluate the overall modeling task. Thin wall molding has been selected to keep the weight of the product to a minimum, and to provide the maximum amount of space internally in the unit for the circuit pack and LCD screen.

- There is a flat parting surface or plane between the housings.

- The larger flat surface will have a wall thickness of 1 mm and the side walls will have a wall thickness of 1 mm. Therefore, a constant wall thickness of 1 mm will apply.

- There is one mounting screw at the back of the unit. Therefore, alignment ribs must be used to guide the unit together properly.

- The large, flat surface will be slightly crowned to avoid deformation during the part cooling cycle and to help add rigidity to the assembly.
- There is a texture applied to the external surface for cosmetic purposes.

> ✓ **TIP:** *When working on a design of this nature, avoid designing features into your model that are likely to cause flash in the mold. This is a good idea in all plastic design but is particularly critical in a thin wall design due to the increased pressure in the mold cavity.*

The mold itself will require polishing both internally and externally to ensure that the part comes free from the mold easily during the processing cycle. The following section takes you through the process of modeling the pager component.

Modeling Process

EXERCISE

Phase 1: Create the Basic Shape

Starting with a protrusion 50 mm by 50 mm, extrude the protrusion 10 mm from the parting plane. Create an advanced variable section swept cut applied to the external surface by sweeping a 200-mm arc along the path shown in the following illustration. This trajectory is sketched on the center datum plane. Use a 3-point arc when sketching the trajectory on the center datum plane. Set the bottom datum plane to be the pivot direction plane.

Cut trajectory.

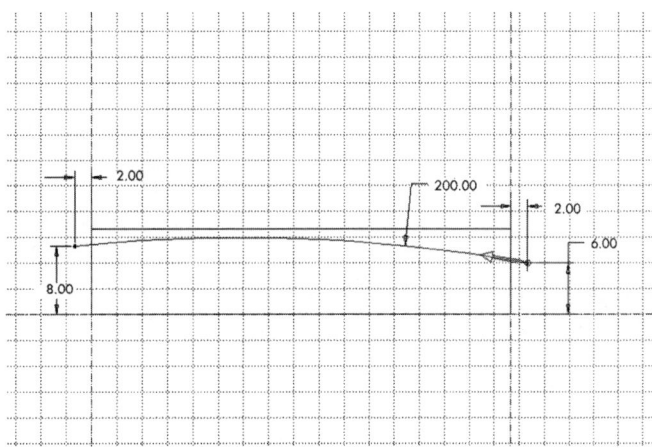

The sketch for the cut is a radius of 150 mm and follows a single trajectory, shown in the previous illustration. The entire feature is created as an advanced variable section sweep with the pivot direction applied to the parting plane. You can take advantage of a sketcher point to hold the tangency of the radius on the path of the trajectory, as shown by the point in the first of the following illustrations. The second of the following illustrations shows the finished crown.

Sketch of the cut profile.

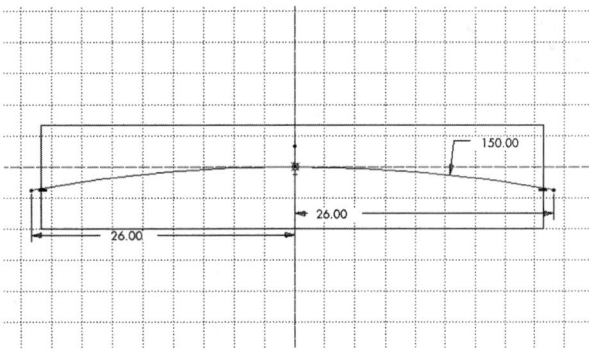

➽ **NOTE:** *The previous illustration shows a sketch point at the trajectory center to hold the arc tangent at that point.*

Finished crown on external surface.

Phase 2: Add Draft and External Rounds

Draft may be added to the external surfaces (sides) of the part at a value of 4 degrees using the parting plane as the draft reference plane. Finishing rounds of 5 mm are then added to the two bottom corners, and then a finishing round is added at a value of 3 mm all around the part except for the lens end and parting surface. The following illustration shows the features modeled in this phase: draft on four surfaces, two edges rounded at 5 mm, and a chain round with a value of 3 mm.

Draft and rounds added externally.

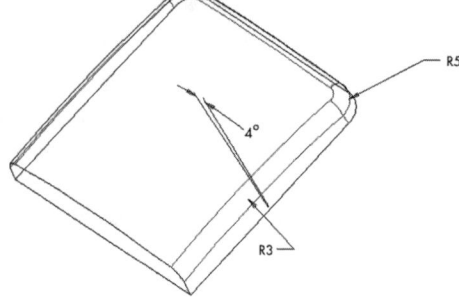

With the external rounds and draft in place, you can proceed with the shell creating the internal geometry.

Phase 3: Create the Shell

Select the end surface where the lens is located, and the inner surface, to be removed by the shelling process. This will open the end of the unit for the lens window, as well as provide a constant wall thickness throughout the part. The first of the following illustrations shows the surfaces to be removed by the shelling process. The second of the following illustrations shows the finished shelled geometry.

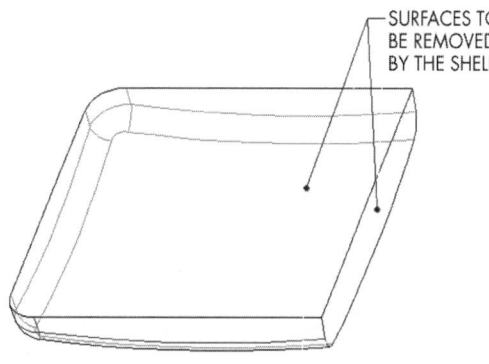

Surfaces selected for removal by the shelling process.

Finished shelled geometry.

With the geometry shelled, you can now start modeling the internal geometry of the component.

Phase 4: Create Interlock Geometry

The geometry used to join two housings may be modeled using the traditional plastic interlock. Much like the project in Chapter 8, ribs may be added that serve as guides during assembly and that stiffen the joint between the two housings. The interlock rib must act on a wall thickness of 1 mm. This does not leave much room for the seating surface between the two components, as the surface must be offset by 0.5 mm to provide the step. Interlock ribs added to the joined parts will act as stiffeners between the two components. The following illustration shows an interlock rib.

✒ **NOTE:** *The top part of the interlock stiffening rib also incorporates a lead-in angle to aid in aligning the mating component at the time of assembly.*

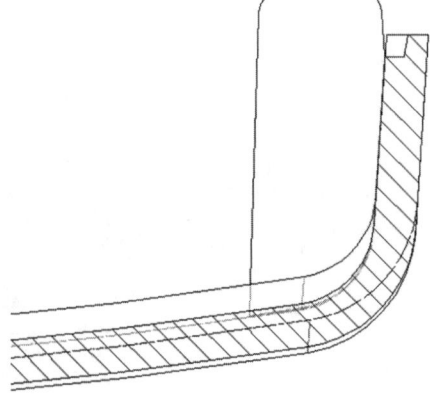

Cross section of the interlock showing an interlock rib.

Modeling two interlock ribs on each side and two at the bottom of the unit should suffice to firmly join the housings. The interlock itself would typically be modeled as a lip feature. The issue for this design is whether or not to have the interlock feature travel all around the joined housing or to employ an intermittent interlock.

There is no need to run the interlock all the way to the end of the part, exposing the interlock detail. There would also be concern over having the interlock travel around the corners of the housing, which would create a very thin wall at a critical location. If the interlock were avoided on the corners, the corners would stand a much better chance of not breaking if the unit were dropped on a corner.

If the interlock were modeled to stop short of the end of the unit, maintaining a 1-mm thickness on the corners, the result would be the component shown in the following illustration. In this example, the interlock ribs were 0.6 mm thick prior to applying draft on their sides. The interlock detail was created using cuts instead of the lip feature.

Finished interlock detail.

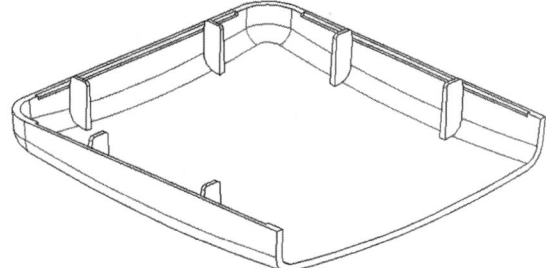

Phase 5: Add Internal Ribs and Screw Boss

The internal ribs would then be added to the design to locate the lens window of the product. This series of two ribs spaced 1 mm apart would be made as protrusions 0.8 mm thick and have draft added to them. It is important to note that 1 mm is the size of a rib at its base.

A screw boss may then be added at the center location of the base using the rules from the part clamping exercise, in which the screw seat is the last component to be clamped together during assembly. Internal ribs for holding the circuit pack may be modeled, as well as any stiffening ribs required for making the unit more robust.

Finishing radii then need to be applied to the internal ribs to complete the model. You may choose to add stiffening ribs in areas where there are large thin patches of geometry. These will also serve as flow runners in the molding process. These features may be left until the first parts are taken off the tooling. They can be placed in the critical areas at that time. The added ribs, draft, final radii, and screw boss are shown in the following illustration.

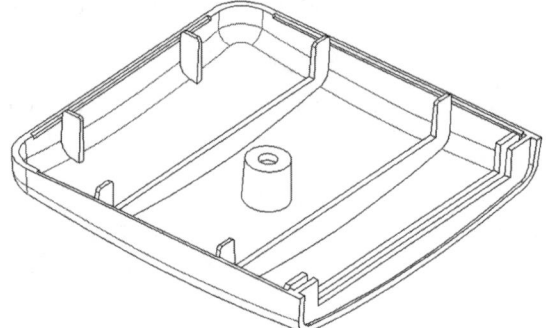

Internal ribs, draft, and final radii added, and screw boss located.

With the modeling complete, you can check the part for a common wall thickness to ensure that the component does in fact maintain a thin wall of 1 mm.

Phase 6: Check the Part

When the design is complete, or at any time during the design cycle, you can check the component to ensure the wall thickness is being maintained and that you have not modeled something that is unmanufacturable. The following are a number of points to consider when evaluating geometry.

Modeling Plan and Process

- Using the Info Draft Check tool and setting the angle to be checked to that of the minimum draft angle, you can make sure adequate draft has been applied to all surfaces of the component. Using Draft Check, set the draft angle to 2 degrees and use the parting plane to evaluate the part as the draft surface selected. Draft Check will provide a color display of the surfaces of the solid model that you can analyze visually, as shown in the following illustration.

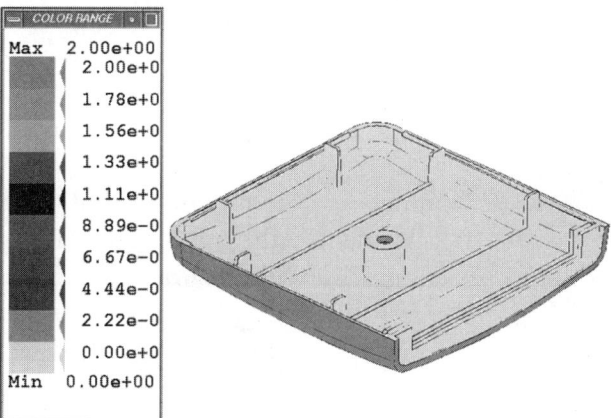

Draft Check tool applied.

- Cross-sectional views can be used to quickly visualize a section of the solid model. Consider using a make datum and create a planar cross-sectional view that may be modified to relocate itself elsewhere in the part. You are able to modify the section location to view any area in the part. Remember to create the cross-sectional view in both directions of the part so that you can examine the wall thickness thoroughly. The following illustration shows a cross-sectional view.

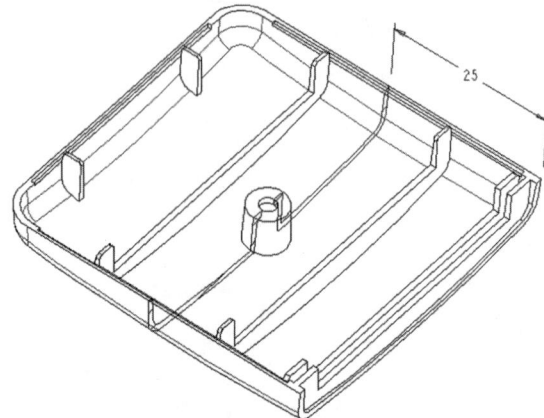

Cross-sectional view.

- You may use the minimum wall thickness tool found in the Info section to identify the minimum wall thickness of a component cross section. This will find areas where you may have made the material too thin. For the project presented in this chapter, it will inevitably identify the interlock area as the thinnest wall. This tool allows slices across the component, as shown in the following illustration.

Minimum wall thickness tool.

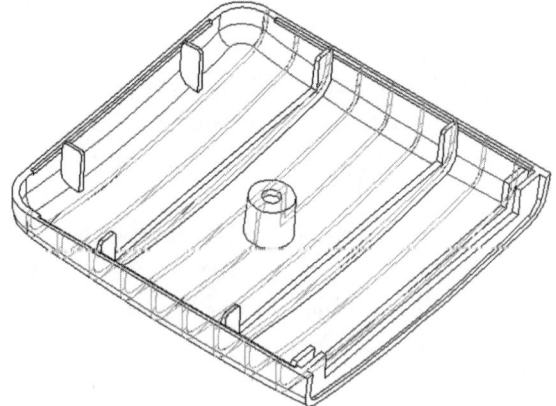

Summary

This chapter has discussed thin wall molding in the context of thin wall technology. Emphasis was placed on determining whether a design is going to use thin wall technology or conventional molding technology, designing for thinner walls than is likely to be the norm at your company. Your plastic suppliers may help you determine which technology would apply to design your projects. Some of the basic rules were covered and how they apply to this type of design. A pager housing was designed using thin wall technology with consideration given to the strength of the product assembly. Tools to analyze your designs were covered to ensure your project would be manufacturable upon completion.

Chapter 12
Extruding Shapes

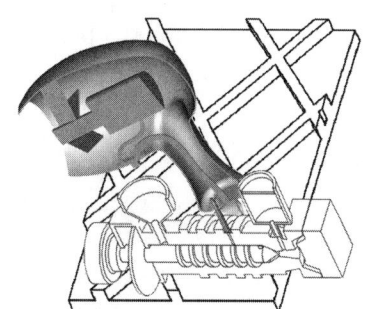

Introduction

Extruded shapes are made from a die as opposed to injection molded parts, which come from a mold having a core and cavity. The die is a cavity, which has a fixed shape. When a cut or a cross section is made perpendicular to the flow of the plastic, through the die, it is always the same, or constant. The plastic is continuously fed through the die to form a long, constant cross section shaped product. Many benefits are obtained by using plastics for extruded parts. The following are some of the benefits.

- The shape can be customized to fit the customer's need. For example, when frames are used to hold glass windows in place, the Pro/ENGINEER 3D solid frame cross section pattern can easily be modified to make similar shapes of dies for larger or smaller windows.

- The plastic extruded piece can be made to have a unique shape and size. Plastic parts can be designed to make shapes that could not be made if metal forms were used. For example, if an extruded support structure was needed to hold and surround coaxial cables (such as those used for cable TV) for protection, either a metal or plastic extrusion could be used.

However, if more coaxial cables were to be added at a later time, the plastic extrusion could be made so that it would temporarily open to allow the new cable to be placed inside the extrusion. A metal extrusion would not be able to be opened and closed. An example of such an extrusion is shown later in this chapter. The thickness of the plastics can vary greatly in an extrusion. The designer can thus make the extrusion with strong, thick sections and finely detailed functioning features in one part.

- The selected plastic can resist most caustic chemicals. The plastic can be selected to resist deterioration by chemicals, which would readily react with metal parts.
- Plastic extrusions are lightweight. This makes the parts easier to handle, lighter and cheaper to ship, and easier to assemble than metal extrusions.
- Plastic parts can be fire retardant and self-extinguishing.
- Plastic is recyclable and reusable.
- The plastic can be any color. It does not have to be painted, which would represent an extra manufacturing cost.

Now that the advantages of extrusion molding have been listed, the designer must design the part so that these advantages can be used to the fullest extent. The methods described in the earlier chapters, such as making a part stronger and preventing warpage, must be applied in this method as well. The following examples display shapes that can be created but must be designed with plastic extrusion and constant shape, and be able to make the part strong without adding cross-sectional ribs or domed features.

Avoiding Warpage

An extruded part should have walls that are constant in thickness. Any extra thick wall will cool at a slower rate and cause the part to warp. Having constant wall thickness, a part will shrink and cool at the same rate everywhere and not warp. The following is an example of a good design for a gasket seal. Note that the walls are the same thickness and that the appearance to the consumer is appealing.

An extruded gasket seal.

Plastics for Support Structures

Plastics are increasingly being used as support structures. For example, pallets traditionally made of wood are now also made of plastic. The following illustration shows a plastic pallet that replaces the wood and is lighter and stronger than wood. The designer makes the part hollow in the middle to reduce weight. The plastic can also be fiber filled to increase its strength.

A supporting surface.

Post Processing on Plastic Extrusions

The final shape of a plastic extruded part can be different from the shape of the part as it comes out of the mold. Post processing, such as moving the plastic to go from an open shape to an enclosed or partially enclosed shape, can be performed on the plastic just after it comes out of the mold and before the plastic becomes a solid. The purpose of creating a different shape from the final shape is to design a shape that can be modified to create a stable and strong final product. The following illustration shows a part that has a cavity to hold cables. The designer creates the part as shown.

Shape of a part as plastic exits the die.

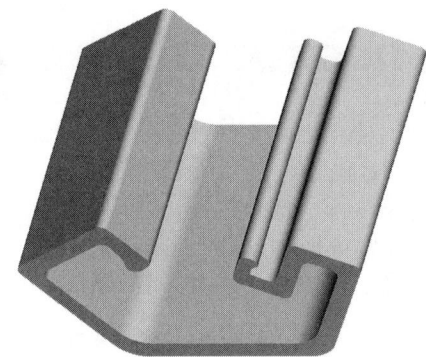

The part can then be pushed and turned to form the final shape, as shown in the following illustration. The part will stay in this shape and have a force on the snapping section. Some initial cooling on the surface of the extrusion will cause an internal stress that will make the part want to open. However, the snap fit will keep it closed. The Pro/ENGINEER designer can use the Measure Thickness tool to determine the thickness of the plastic in the extrusion. This allows the designer to make parts the appropriate thickness so that they will turn in the correct location.

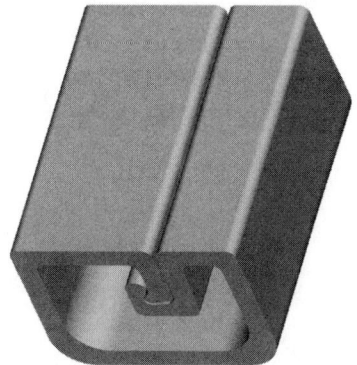

Final shape of the extrusion.

Notice that the part, shown in the previous illustration as snapped together, can be pulled apart if more cable needed to be placed in the cavity section of the extrusion. The final shape could not be made directly from an extrusion die because of the snap fit requirement.

Summary

Extrusions will require the Pro/ENGINEER function Measure Thickness to help in the design of extruded parts. The designer can modify the thickness to adjust for the strength and stress bending of the final part in situations that preclude the possibility of making domed surfaces or ribs to perform this function.

Chapter 13
Blow Mold Designs

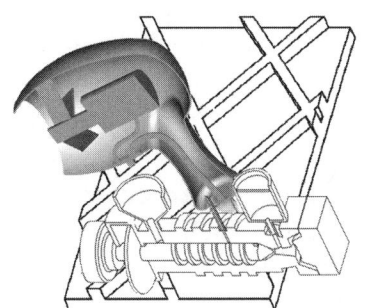

Introduction

Blow molding is different from extrusion and injection molding in that the final wall thickness is not constant. A gas blows the plastic from a vertical hanging tube onto the cavity of a mold, as was described in Chapter 1. The plastic can sag and stretch before it finally cools on the cavity of the mold. The designer must be aware of this fact and design the part accordingly. The designer must also remember that the cavity needs to split into two sections.

Undercuts, an undesirable feature in the manufacture of blow molded parts, are another consideration for the designer. The manufacturer cannot guarantee that the part can be made if slides are required for undercuts. For example, slides are often used to produce the base of a bottle design. Such slides, discussed in this chapter, are easy to design and manufacture.

In this chapter, you will be shown how the features in Pro/ENGINEER can help in the design of blow molded parts. Many large parts, such as water tanks, to small parts, such as toys, can be made using the blow molding process. The example presented in this chapter is of a simple bottle design, demonstrating the advantages of 3D design.

Designing a Bottle for Blow Molding

To design a bottle, you need to know the required volume, height, width, handle size (if a handle is required), and any special cosmetic features. Once these basic requirements are defined, the product may be designed. The example discussed in this chapter is the bottle with threaded neck shown in the following illustration. The design task described in the sections that follow will begin with the neck of the bottle and proceed to the basic shape of the bottle, the top, labeling, the bottom, and the volume (body). These tasks arre described to show the progression, from the most difficult to the least difficult features, in making a part. A design procedure for creating the features of the product, from the critical shape to the least critical shape, is described.

Bottle with threaded neck.

Neck Section

The neck section of a bottle, shown in the previous illustration, is the area between the main section and the cap section. In this section, the curvature must be smooth and tangent to the main section and the cap section. Otherwise, there will be lines that show the discontinuity between the three sections. The cap section must be round in order for the cap to fit and screw on.

The neck section is usually the best place for the designer to start because of the constraints previously mentioned. Many options for creating the neck are available in Pro/ENGINEER. These include blends, solid protrusions, trajectories, and surfacing. The choice of methods depends on the intended shape of the final product. The designer should choose those methods that incorporate the greatest flexibility should the shape need to be changed for volume or overall appearance—a principle that applies to the majority of design tasks.

In the case of the bottle, you will use a Pro/ENGINEER solid protrusion blend method. The blend function employs sketches that describe the shape of the part as it changes from the base section sketch to the top section sketch. Each section (sketch) created and changed by the designer describes the profile of the shape needed to provide the correct flow from the top to the bottom of the blend. The designer may make as many sketches as necessary to create a described shape.

The change from sketch to sketch is called toggling. Each of the sketches must have the same number of curves for the blend feature to function; otherwise, the final shape cannot be recognized properly. The first and second sketches have four lines with fillets joining the lines. Two of the lines are large radius curves, which ensure that the part can be removed from the mold at the parting line. The last sketch shows a circle intersected by lines to create eight equal-length curves. The following illustration shows the result of using the blend function.

Bottle neck section.

At this point, there is a problem. At the top and bottom of the bottle, the walls are not perpendicular to the base. To change the top section of the neck, a revolved protrusion is added. This removes the requirement for the perpendicular surface and enhances appearance. At the bottom wall, the edges are used to extrude a solid shape at the desired depth. The method for creating these features is described in the section that follows.

The Basic Shape

The top of the bottle must be made round so that a cap and threads can be placed on it. A protrusion is added to the top of the neck section. This protrusion is smaller than the neck section so that the cap will stop at this point. Rounds are added to the area where the base and the neck section meet. Rounds are also added to the bottom of the bottle. The following illustration shows a basic bottle shape.

Basic bottle shape.

Top Section

The angle at the top of the bottle's neck should be perpendicular to the base so that you can design a Pro/ENGINEER helical feature with which to create a screw thread. The screw feature (threads) for the cap can be added easily to the part. Once the top part is added, the screw can be completed. In Pro/ENGINEER, there is a function called Helical Sweep that allows the designer to make various helical shapes with differing pitches and angles. The function is also used to create right- and left-handed screws.

You make the helical sweep using two sketches found in the advanced feature option for solids. The first sketch determines the length and angle of the sweep. The second sketch makes the profile of the sweep, such as the shape of the screw threads for the bottle. You can change the shape, pitch, and direction of the sweep, which makes it possible to create a left- or right-handed screw thread. The ends of the threads can then be shaped to the desired contour so that they will smoothly join the neck of the bottle. The following illustration shows the solid model of the bottle.

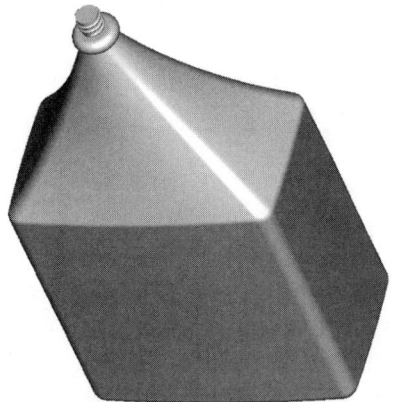

Solid model of bottle.

A second method of modeling the helical threads is to make a datum curve using the equation option. Using the cylindrical coordinate system, the Pro/ENGINEER designer might employ a formula such as the following.

```
Radius = Pitch * Angle
```

Angle determines the number of times the helical curve goes around, and Pitch determines the tightness of the helical curve. It is up to the designer to decide which method is easiest to employ. The next step in the design process is to add labels and special features to the product.

Labeling

Pro/ENGINEER includes a function called Tweak Offset. This function allows you to offset features from a surface of a part with a positive or negative offset. This functionality will be used to add the word Soap to the face of the bottle, as shown in the following illustration. An offset datum plane is used as a sketching plane, where the letters S, O, A, and P are sketched. This sketch is then projected onto the bottle with a positive offset. The offset is made normal to the sketching plane in order to make the part removable from the mold.

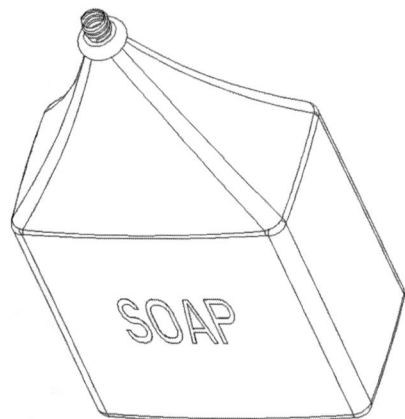

Bottle with label.

The advanced section in the datum curve menus can also be used to create words without sketching the letters. The letters are automatically created and are automatically used later with the tweak offset to create the labels on the bottle.

Bottom

The bottom of a bottle is typically not flat but rests on round edges with a slight indentation in the base of the bottle. The indentation exists for two reasons. One is to strengthen the bottom of the bottle to prevent the weight of the liquid from bending it out. The second purpose of this shape is to prevent the bottle from forming a vacuum between the bottle and a wet surface, causing it to stick to the surface and increasing the likelihood of spills.

Adding an indentation and rounds means that the mold will require two inserts, as described in Chapter 1. Such inserts are easily manufactured and easily incorporated into the mold, adding insignificant cost to production of the mold. The indentation is created using a solid cut, with rounds incorporated. The bottom of the bottle is shown in the following illustration.

Bottle bottom.

Shell and Volume

The next step is to shell the bottle to the desired thickness. Once the bottle is designed, the volume within the bottle can be determined. In this phase of the process, the designer places a datum surface at the desired height for fluid in the bottle. A quilted surface shape representing the inside volume is created by selecting the inside surfaces of the bottle, merging these surfaces, and then using the datum surface as an intersection plane, described in Chapter 15.

The quilted surfaces are then transferred to a new part and made into the solid feature. The volume of the fluid can then be displayed from the Info section and the mass property function. The shape of the model and the location of the fluid level datum plane can be modified until the desired volume is achieved.

> **NOTE:** See Chapter 15 for a more detailed explanation of volume measurements.

Summary

The use of blow molding in plastics design is most common in the creation of bottles or objects that require hollow middles. Pro/ENGINEER offers the designer tools for creating the shapes appropriate to blow molding and for checking the volume of products made with this process. As evident in the bottle example presented in this chapter, the designer can employ numerous methods for creating the features of parts and products intended for the blow molding process. The threaded (helical) top of the bottle is an example of such features. When creating such features, keep the following points in mind.

- Use helical shapes for the tops of containers.
- Plan for continuity between the cap or top of the container and the main section of the container.
- Make the base of the container strong; for example, by adding inserts.

Chapter 14

Mirror Parts

Mirror Image Part Creation

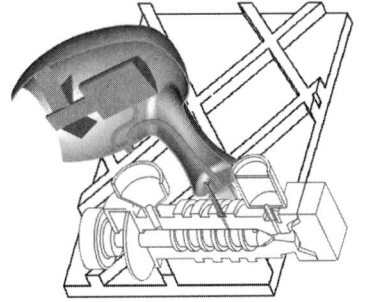

Introduction

Pro/ENGINEER provides a utility within the assembly module that enables you to create mirror image parts with ease. This chapter explores procedures for creating mirror image parts in your designs. The information is based on making individual components that are mirror images of each other, which should not be confused with mirroring geometry. Mirroring geometry refers to modeling 50 percent of the geometry for a part and mirroring the rest across a datum plane to achieve 100 percent of the part.

A Mirroring Project

The project selected for this chapter entails modeling a left-hand version of a part and using Pro/ENGINEER's assembly tools to mirror the component to produce a right-hand version of the part that is an independent Pro/ENGI-NEER model. This technique is common when creating mirrored components and is very expedient.

Project Description

The project that follows highlights the issues involved in mirroring duplicates of parts. The project involves a latch system for which the left-hand latch has been previously modeled and the designer wishes to achieve the right-hand version from the existing model. The following illustration shows the left- and right-hand versions of the latch assembled to the lower cover of a plastic box.

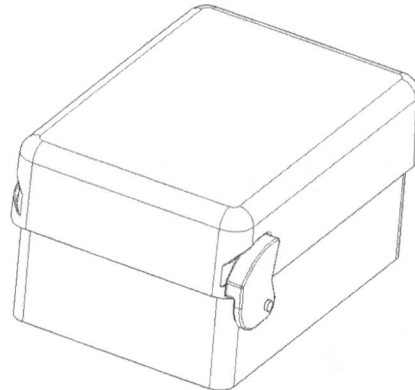

Assembled latch mechanism and housing.

The intent of this project is to get the left-hand part ready, add it to an assembly, and then create the right-hand version. Any geometry exclusive to one of either latch should be omitted from the model until after the mirroring has been completed. The following section presents the phases involved in modeling and mirroring the latch.

Latch Modeling Process

EXERCISE

Phase 1: Prepare Initial Part for Mirroring

If you consider the left-hand latch a stand-alone model with which to begin the modeling process, you need to add some information to that model to determine the location for mirroring. That is, you need to model the plane at which the mirroring will take place. This can be

A Mirroring Project

done in the part or in the assembly that incorporates the part. Because the information important to the part should travel with the part file, you should add the plane to the part file if for no other reason than as a reference. If there are any other features that would aid in assembling this model, they should also be added to the part file.

The following illustration shows the left-hand part model with the mirror plane added. This enables you to add the component to an assembly and center the part for the creation of the mirrored part. Using this technique, the part will be both modeled and ready for assembly in its final position.

Center mirror plane added to the left-hand part model.

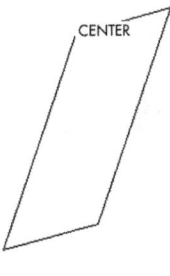
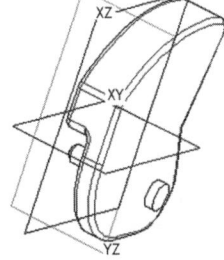

Phase 2: Add Left-hand Model to an Assembly

If an assembly were created at this point, the left-hand model could be added to the assembly and the right-hand part created. An existing assembly could also be used to accomplish this task. For the purpose of clarity, you will create a new assembly. Incorporate the left-hand model into the assembly such that the mirroring plane is the central plane.

Phase 3: Create the Right-hand Model

In the active Pro/ENGINEER assembly, select Modify Component from the menu. Select Mirror and create the new component by mirroring about the central datum plane. The resultant geometry will give you a right-hand version of the left-hand model (mirrored copy), which will be an independent Pro/ENGINEER part file. If further detail is required on one of the latches, it may be added at this time. The following illustration shows the assembly with the left- and right-hand models included.

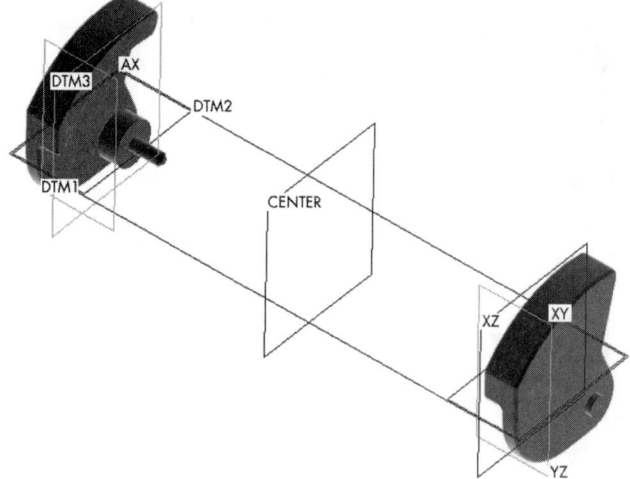

Left- and right-hand models shown in the assembly.

It is important to note that because the models are created independently, changes to one model will not be automatically made to the other. The following illustration shows the right-hand component as an independent Pro/ENGINEER model.

Summary

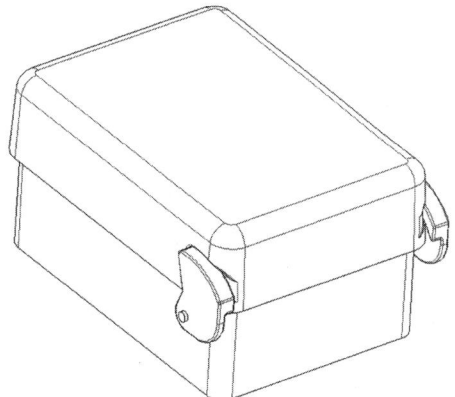

Finished right-hand model.

Summary

Mirror image parts may be created with ease using Pro/ENGINEER mirroring tools in the assembly module. Consideration should be given to whether the parts are exact duplicates of each other prior to modeling the geometry to be mirrored. The latches chosen for the project presented in this chapter were mirrored in Pro/ENGINEER to form a left- and right-hand version of the geometry, saving considerable modeling time.

It is important to remember to use this technique to create independent models of geometry when you need to mirror a complete part. Consideration may also be given to the advantages of design changes after the first parts are issued. You might find that if design changes are extensive enough it is easier to change the left-hand part and repeat the mirroring process rather than change both components.

Chapter 15

Volume Calculations

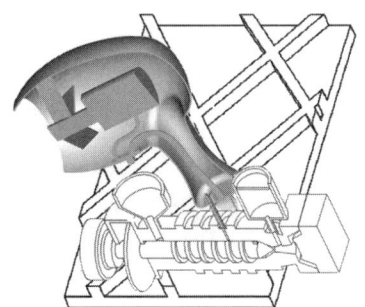

Introduction

Pro/ENGINEER gives you the ability to perform engineering calculations on solid models. Proper understanding of the calculations and how to use them helps you create designs of higher quality. The basic concepts and first principles of the properties are not developed in this chapter. The intention is to show you how to use and create the correct engineering values to arrive at competent and reliable results every time. Your goal as a designer is to be confident that the values you implement agree with the dimensions used in the design of any given plastic part. The section that follows will give you confidence that the values you use will be correct for your engineering analysis.

Calculating Volumes

Two uses for volume calculations can be applied to designing products and using Pro/ENGINEER. One is to find the volume of plastic necessary to the manufacture of a plastic part. The second is to determine the volume of space left inside a plastic part for which this is required. That is, volume calculations are used to determine how much space is available for fluids (including air) or solids. For example, the volume calculations for the space in a plastic loud speaker cabinet are critical to the design, and therefore the performance, of the speaker. A volume calculation is also critical to the proper design of such things as bottles or other types of containers intended to hold a liquid.

> **NOTE:** *See Chapter 13 for initial discussion of designing for fluids.*

Volume of Plastic in a Part

For Pro/ENGINEER to determine the volume of plastic in a part, the part must be completely solid, and the units of measurement must be defined. You define the units for length and mass in the Set up section. It is very important to keep units consistent for all parts associated with a given product. This helps guarantee that the engineering values, such as the moment of inertia, will be correct and will make sense to the designer.

The Info Mass Props option shows the volume and surface area of the part. The volume tells you the minimum volume of plastic needed to make the part. The surface area gives the total surface area of the part. The surface area important to a plastics designer is the maximum cross-sectional area parallel to the parting line. The larger the cross-sectional area, the larger the clamps need to be on the molding machine to hold the mold together as the plastic is injected. Tables of relations can be obtained from mold manufacturers. These show the relation of surface area of a part to size of molding machine required to make the part. This information helps the designer estimate the production and material costs of a plastic part.

Volume Within a Plastic Part

A plastic bottle serves as a good means of learning how to calculate the volume inside a container. To do this, you would create a surface plane at the proposed height of the liquid in the bottle, as shown in the following illustration. The surface is used as a cutting plane to remove the top portion of the bottle. The remaining shape represents the volume of material in the empty bottle. The mass property table is used to show the volume of the empty bottle.

Calculating Volumes

Volume in a bottle.

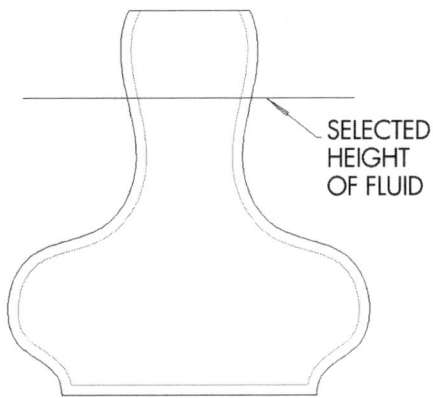

To find the liquid volume inside a bottle, the volume must be made a solid, as previously described. Because the volume required is a liquid, you need a method of creating a solid equivalent of the liquid volume. To do this, you first select all inner surfaces of the object. The purpose is to eventually join all of the surfaces to form a solid part that describes the inside volume of the part.

In the case of the bottle, there are two interior surfaces, as indicated by the lines in the following illustration. Although this part was made from a revolving protrusion of 360 degrees, Pro/ENGINEER divides the circle into two arcs. Keep this in mind and make sure to pick all interior surfaces of the object for which you are trying to calculate a volume.

Two inside surfaces of a bottle.

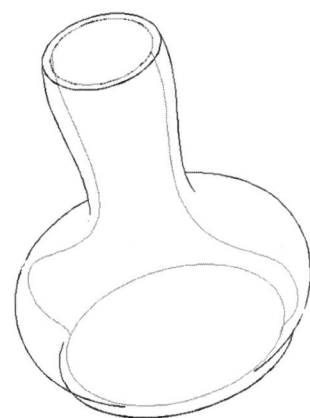

The designer uses the feature for joining merged surfaces to join all surfaces within the bottle. Now the designer has two surfaces: the flat, extruded surface that describes the top level of the fluid and the inside surface of the bottle. The next step is to intersect these two surfaces. The designer uses the Intersect Merge feature and selects the two surfaces. Pro/ENGINEER finds the intersection of the two surfaces and displays this intersection line, as well as the first surface selected. In this case, the flat surface was selected, and is therefore the surface shown. The following illustration shows the flat, extruded surface with two arrows pointing to the center of the bottle.

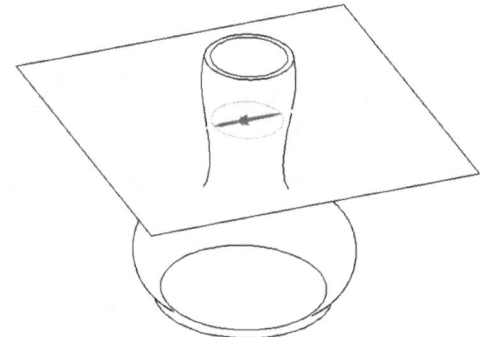

Selecting the correct side of the flat plane.

In the previous illustration, the arrows are slightly overlapped. By flipping the arrows back and forth, the designer can be sure that the correct surface segment is being saved. In the case of the bottle, you want to keep the inside of the intersection of the plane because this will give you the top part of the volume of liquid. This selection would then be accepted.

Pro/ENGINEER automatically shows you the other surface, which in this exercise is the inside surface of the bottle. The arrows, at the intersection line of the two surfaces, are shown to determine which section of the inner surface to keep. The following illustration shows the arrows and the correct direction for selecting the inside surface.

Calculating Volumes

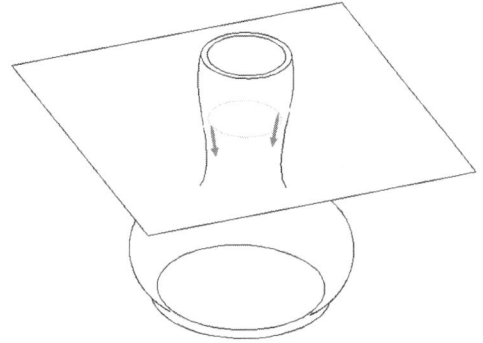

Selecting the correct side of the inside surface of the bottle.

Again, the designer can flip the direction of the arrows to ensure that the correct surface is being kept. This feature is extremely important when you are intersecting complex surfaces. The correct direction is selected and the final enclosed surface is created, as shown in the following illustration.

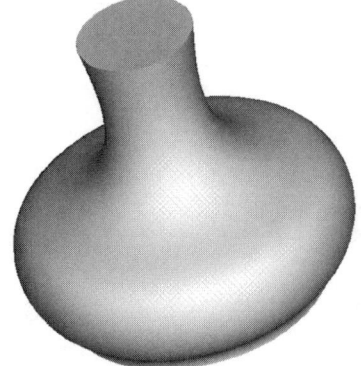

Inside surfaces of the bottle.

This surface model is a completely enclosed, watertight model. The enclosed surface must be made into a solid to enable Pro/ENGINEER to determine volume within the surface. Using the solid protrusion feature with the quilt option, the surface can be made into a solid. However, because the surface has two sides, an inside and outside, Pro/ENGINEER does not know which side to use to create the solid. The arrow shown in the following illustration helps the designer determine which side of the surface should be made solid. Once again there is the option of flipping back and forth to be sure that the correct side of the surface is being selected.

Selecting the inside volume of a surface to create a solid.

The part is now completely solid and the total volume can be found in the mass property table. By subtracting the original volume of the empty bottle (after using the surface plane to remove the top portion of the bottle) from the new volume of the full bottle, the inside volume is found. This method can be used for any shaped part to determine the volume inside the part.

Summary

Using Pro/ENGINEER, you can determine the volume of a part or the volume of the space inside a part. This can be done for any shape or volume. Because such calculations are very accurate, leaving no doubt in the designer's mind about volumes within containers, this information can be used for the development and costing of products.

Chapter 16

Applying Material and Mass Properties

Introduction

Material property files may be applied to your Pro/ENGINEER models to enable you to perform accurate engineering calculations within Pro/ENGINEER. The material file is found under the Setup option of the PART menu. The terms and definitions found in the material file are described in the following section.

Plastic design work involves various engineering values. Therefore, for the terms defined in the following section, three sample units are provided. The sample units are from the gram/millimeter/second, metric (System International, kg/m/s), and English (lb./in/s) systems.

The Material File

The material file contains a list of parameters that define the material. The parameters and sample numbers are defined in the list of terms that follows. The sample numbers used are typical of a plastic material. The numbers change for each plastic material and can be obtained from the resin supplier. These numbers are used to display the general magnitude that should be used for plastic materials, as opposed to magnitudes for metals. These numbers are determined at room temperature.

Material File Terms

The following are the material file terms with which you should be familiar.

Young's modulus. The elastic modulus, or modulus of elasticity. It is the constant of proportionality between the stress and strain of the material. Examples of values of stress: 2,300,000,000 g/mm/s/s; 2,300,000,000 pa.; 330,000 psi.

Poisson's ratio. In all solid materials, deformation occurs in the direction of the applied force (axial) and in the direction at right angles (lateral) to the applied force. Poisson's ratio is the ratio of the lateral strain to the axial strain. It is used only in the elastic region of deformation. It is a unit-less number and can be set to 0.3.

Shear modulus. This modulus is used only for small deformations and in the elastic region. It is the shearing modulus of elasticity and is the constant of proportionality between the shear stress and shear strain of the material. Examples of values of stress: 2,600,000,000 g/mm/s/s; 2,600,000,000 pa.; 360,000 psi.

Mass density. The mass of a material, not its weight. The values are 0.0000384 grams per cubic millimeter, or 38.4 kilograms per cubic meter, or 0.00139 pounds mass per cubic inch.

Thermal expansion coefficient. A material expands and contracts as the temperature increases or decreases. The ratio of the change in length per degree of temperature to the original length is called the thermal expansion coefficient. Plastic materials expand at different rates due to thermal expansion. The designer must be aware of the expansion of material in the design to avoid sloppiness or rattling of the part as the temperature changes and parts made of different plastics are no longer in complete contact with each other. The real values are 0.000095 millimeters per millimeter per degree Celsius, or 0.000095 meters per meter per degree Celsius, or 0.000053 inches per inch per degree Fahrenheit. The values that should be placed in the Pro/ENGINEER material file should be 9.5, 9.5, and 5.3, respectively.

Thermal expansion reference temperature. A base reference temperature is required to determine the expansion of a material for finite element analysis. The designer can choose whichever temperature is needed for the analysis. The units will be in Celsius or Fahrenheit degrees.

Structural damping coefficient. A coefficient used in finite element analysis for transient dynamic response for displacement solutions. It is difficult, if not impossible, to find for plastics. The designer would have to work with the plastic resin company to determine the values for each resin.

Stress limit for tension. Analyses using these values are only good in the elastic region of the plastic. This value is the maximum stress that can be applied to the plastic before it goes into the plastic region of the deflection. The values are 41,000,000 g/mm/s/s, or 41,000,000 pa., or 6,000 psi.

Stress limit for compression. The maximum stress value for compressing a plastic before it goes into the plastic region. These values can vary greatly, depending on the type of plastic and whether fillers are used. The values are 25,000,000 g/mm/s/s, or 25,000,000 pa., or 3,700 psi.

Stress limit for shear. The maximum shearing stress for a plastic before it goes into the elastic region. The values are 41,000,000 g/mm/s/s, or 41,000,000 pa., or 6,000 psi.

Thermal conductivity. This value represents the ability of the plastic to transfer heat by conduction. The typical values are 0.00021 watts per millimeter per degree Celsius, or 0.21 watts per meter per degree Celsius, or 0.12 BTU per hour per foot per degree Fahrenheit.

Emissivity. The ability of a body to emit radiation over all wavelengths. The emissivity of a material varies with the temperature and wavelength of the radiation. Perfect black-body emissivity would have a value of 1. All materials have values lower than 1, which are found by experimentation. The emissivity values are a ratio of real emissive power and absolute black-body emissive power and are unit-less. The typical value is 0.9.

Specific heat. A coefficient that relates the change of heat put into a solid to its change in temperature. The values are 25.1 watt-second per gram per Celsius degree, or 25,100 watt-second per kilogram per Celsius degree, or 6 BTU per pound mass per degree Fahrenheit.

Hardness. A number that refers to stiffness or temper, or to resistance to scratching, abrasion, or cutting. There are many hardness scales, such as Brinnell, Rockwell, and Vickers. The user has to know which scale is being used before any value can be placed in this feature.

Condition, initial bend y factor, and *bend table.* Terms specific to analysis programs. The values and units must be determined by the user for each of these terms and then be placed in the materials table with the correct units.

Material Files and Finite Element Analysis

Plastic materials change properties as the temperature changes. The designer must be aware of the changes and use the correct values for any finite element analysis performed. The following is an illustration of a material file. The material file can be passed along with the Pro/ENGINEER part file to a finite element solver for computer-aided engineering (CAE). The material properties will have been defined with the proper units, which reduces the chance of incorrect analysis being performed because of improper inputting of material values in the CAE program.

Sample material file.

```
MATERIAL  PLASTIC_SAMPLE

This file may be edited using available editor.
Just type on the necessary lines appropriate values
after the "=" sign. Comments are not permitted on
lines containing material properties names.

YOUNG_MODULUS                    =   2.300000E+09
POISSON_RATIO                    =   3.000000E-01
SHEAR_MODULUS                    =   2.600000E+09
MASS_DENSITY                     =   3.840000E-05
THERMAL_EXPANSION_COEFFICIENT    =   9.500000E+00
THERM_EXPANSION_REF_TEMPERATURE  =   2.000000E+01
STRUCTURAL_DAMPING_COEFFICIENT   =
STRESS_LIMIT_FOR_TENSION         =   4.100000E+07
STRESS_LIMIT_FOR_COMPRESSION     =   2.500000E+07
STRESS_LIMIT_FOR_SHEAR           =   4.100000E+07
THERMAL_CONDUCTIVITY             =   2.100000E-04
EMISSIVITY                       =   9.000000E-01
SPECIFIC_HEAT                    =   2.510000E+01
HARDNESS                         =
CONDITION                        =
INITIAL_BEND_Y_FACTOR            =
BEND_TABLE                       =
```

The Mass Properties File

The mass properties file is found in the Info section of Pro/ENGINEER. It contains information helpful to the plastics designer. The mass, inertia, and rotation of a body is important for the design of a plastic part. If the part is one that might be empty or filled, such as a watering container, the design should balance the object properly with and without material (such as water) in it. The sections that follow define and describe terms found in the mass property file.

Center of Gravity

The center of gravity of a part is considered the location at which the force of gravity distributed over the volume of the part produces a weight that may be taken as a concentrated force without any moments. Pro/ENGINEER solves the center of gravity and displays it with respect to the coordinate system. The following illustration shows the location of the center of gravity of a simple part.

Center of gravity of a part.

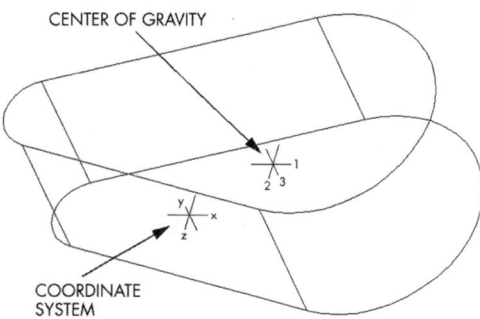

Moment of Inertia

The mass moment of inertia of a body is a measure of the resistance to rotational acceleration due to the mass or inertia of a body. This is similar to a mass of a body being resistant to translational acceleration. The designer can use these values to determine a part's relative resistance to rotation.

Principle Moments of Inertia

A 3D plastic part has three moments of inertia about three mutually perpendicular coordinate axes, and three products of inertia about the three coordinate planes. Pro/ENGINEER uses the products of the inertia of the mass. A plastic part has three moments of inertia about three mutually perpendicular coordinate axes, and three products of inertia about the three coordinate planes.

With an unsymmetrical body and for a given origin of coordinates there is one orientation of axes for which the products of inertia vanish. These axes are called the principal axes of inertia. The moments of inertia about these axes are called the principle moments of inertia. The rotation matrix and angles are used to show the change in orientation of the mass and the principle axis.

Radii of Gyration

Radii of gyration are ratios of the moment of inertia to the mass of the part. They are a measure of the distribution of the mass about the axis. It gives the designer an idea of how the material is distributed around the part and how resistant it is to rotation. The following illustration shows a mass properties file.

Mass properties file.

```
              MASS PROPERTIES OF THE PART TWO
                 VOLUME  = 5.4439081e+06  MM^3
             SURFACE AREA = 1.9482604e+05  MM^2
                 DENSITY = 3.8400000e-05  GRAM / MM^3
                 MASS    = 2.0904607e+02  GRAM

CENTER OF GRAVITY with respect to _TWO coordinate frame:
X   Y   Z    8.7817018e+01  6.0417886e+01  0.0000000e+00  MM

INERTIA with respect to _TWO coordinate frame:  (GRAM * MM^2)

INERTIA TENSOR:
Ixx Ixy Ixz      1.4026244e+06 -1.1091396e+06 -1.1300948e+01
Iyx Iyy Iyz     -1.1091396e+06  3.6280949e+06  0.0000000e+00
Izx Izy Izz     -1.1300948e+01  0.0000000e+00  4.2603646e+06

INERTIA at CENTER OF GRAVITY with respect to _TWO coordinate
frame:
                                              (GRAM * MM^2)
INERTIA TENSOR:
Ixx Ixy Ixz      6.3953913e+05  0.0000000e+00 -5.4155368e+00
Iyx Iyy Iyz      0.0000000e+00  2.0159675e+06  0.0000000e+00
Izx Izy Izz     -5.4155368e+00  0.0000000e+00  1.8851519e+06

         PRINCIPAL MOMENTS OF INERTIA:  (GRAM * MM^2)
I1  I2  I3       6.3953913e+05  1.8851519e+06  2.0159675e+06

ROTATION MATRIX from _TWO orientation to PRINCIPAL AXES:
                 1.00000        0.00000        0.00000
                 0.00000        0.00000       -1.00000
                 0.00000        1.00000        0.00000

ROTATION ANGLES from _TWO orientation to PRINCIPAL AXES
(degrees):
angles about x  y  z   90.000         0.000          0.000

        RADII OF GYRATION with respect to PRINCIPAL AXES:
R1  R2  R3       5.5311133e+01  9.4962507e+01  9.8202096e+01  MM
```

Cross Section Mass Properties

A plastics designer may want to determine engineering values for the section moduli in a part. A cross section can be made in Pro/ENGINEER. This cross section can then be chosen in the Info section with Mass Properties. The cross section is chosen and the engineering values are returned, as shown in the following illustration.

Cross section mass properties file.

```
             MASS PROPERTIES OF THE CROSS SECTION FOUR
                 AREA =  5.3669924e+04 MM^2

       CENTER OF GRAVITY with respect to _FOUR coordinate frame:
       X   Y                    2.0055361e+02  8.3420436e-02   MM

       INERTIA with respect to _FOUR coordinate frame:   (MM^4)

       INERTIA TENSOR:
       Ixx Ixy              1.4012022e+08  -1.1600822e+06
       Iyx Iyy             -1.1600822e+06   2.5775968e+09

       POLAR MOMENT OF INERTIA:             2.7177171e+09 MM^4

       INERTIA at CENTER OF GRAVITY with respect to _FOUR coordinate
       frame:
                                            (MM^4)
       INERTIA TENSOR:
       Ixx Ixy              1.4011985e+08  -2.6216992e+05
       Iyx Iyy             -2.6216992e+05   4.1889847e+08

       AREA MOMENTS OF INERTIA with respect to PRINCIPAL AXES:(MM^4)
       I1 I2                1.4011960e+08   4.1889872e+08

       POLAR MOMENT OF INERTIA:             5.5901832e+08 MM^4

       ROTATION MATRIX from _FOUR orientation to PRINCIPAL AXES:
                             1.00000        -0.00094
                             0.00094         1.00000

       ROTATION ANGLE from _FOUR orientation to PRINCIPAL AXES
       (degrees):
       about z axis                           0.054

       RADII OF GYRATION with respect to PRINCIPAL AXES:
       R1 R2                5.1095653e+01   8.8346434e+01   MM

              SECTION MODULI and corresponding points:
                        MODULUS              1             2    COORD
       about AXIS 1:  1.27013e+06 MM^3    7.5786e+01  -1.1032e+02 MM
                      1.25591e+06 MM^3    5.0918e+01   1.1157e+02 MM
       about AXIS 2:  2.08884e+06 MM^3   -2.0054e+02   1.0547e-01 MM
                      2.74587e+06 MM^3    1.5256e+02  -2.4462e+00 MM
```

The designer can use this as a local, in-depth study of the part. Changes may be made to the part if the inertia or center of gravity in the sections do not meet design specifications.

Summary

The designer has the tools to determine the mass and volume and other engineering parameters to aid in the design of plastic parts. These calculations would be almost impossible to perform were a 2D design tool used. This is where Pro/ENGINEER can immensely aid the plastics designer in creating a well-engineered part. In addition, with the material file the designer can pass a part on to the analytical tool, such as a finite element tool, to perform simulated stress tests.

Chapter 17

Information and Clearance/Interference Tools

Introduction

The information tools used to check and verify your Pro/ENGINEER models may be considered to fall under two main categories. The first category is numerically defined tools that allow you to input or output numerical data into the software. This could be a simple as measuring the length of an existing edge or the distance between two points. The second category is primarily visual. The software conveys information about model curvature, slope, and so on through the use of color-coded displays. The exception to the two categories may be the Draft Check tool, for which you can establish the limits of the software's application when it creates color displays. This means that the tool can display a more specific application in terms of your being able to set the requirements and view the results visually in color.

The third major area covered in this chapter is identifying the clearance and interference in designs involving multiple components. The material covers how to measure clearance between surfaces and presents a number of examples for detecting areas of interference in your product designs.

Most of the visual analysis tools may require user experience to help make a judgment call as to whether the design intent of the component or product has in fact been met. The more the tools are used, the more the user is able to predict in advance of manufacturing the components what the outcome will be.

When you are confirming a finished design in whole or in part, you may find that you will use a number of the analysis tools to check the model structure and stability in manufacturing intent. There is no right or wrong way to use the visual analysis tools, provided you are able to interpret the results in a logical manner. An example of this may be that on a particular component you would need to use the Draft Check tool, referencing more than one reference plane if slides are involved in the part design.

A further method of verifying your designs would be to have the design file manufactured on CNC equipment or a 3D modeling process. This process is difficult to dispute, as you are able to tell if you got what you wanted with the component in your possession. If you do use CNC equipment to verify a finished cosmetic model, ensure that a small cutter is used to make the finished pass, which allows you much more detail in your part's geometric structure.

Note that numerically controlled equipment is expensive. You should attempt to verify your models as much as possible using the software tools available first. Consider cutting a model prior to placing a purchase order for a component to ensure that you do not have poor surface quality, thin sections, or other problems in your finished design.

Numerically Based Information Tools

Pro/ENGINEER's suite of numerically based information tools is used to verify position, tangent conditions, wall thickness, draft angles, and basic mathematical qualities such as length, distance, and radius. These tools can make it easier to verify dimensional relationships and to confirm that geometry has in fact been created as intended. The process of checking your geometry should be employed throughout the modeling cycle, as these tools may be used at any

Numerically Based Information Tools

time. Consider the numerical tools discussed in the sections that follow and how they might be used to verify your designs.

Info Measure

The Info Measure tool is used to verify a model's position and dimensional qualities. This tool may be used to verify the length of edges and curves, and the distance relationships between entities. The tool also allows you to identify the radius of a curved entity at a given point. The use of this part of the tool is relatively straightforward.

Using Datum Points

Datum points may also be placed on a model to provide an entity to measure to. Examples of this are shown in the following two illustrations. In the first illustration, two datum points are placed to identify the extreme width of a product. Once the points are in place, you are able to measure the distance from one point to another. The second illustration shows a component with a curved top. In order to identify the overall height of the part, you would place a datum point on the highest part of the top and measure the height of the component relative to a datum plane at the parting surface.

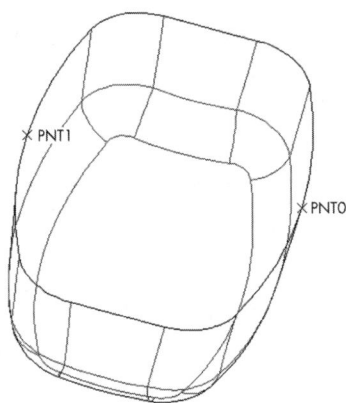

Datum points used to identify component width.

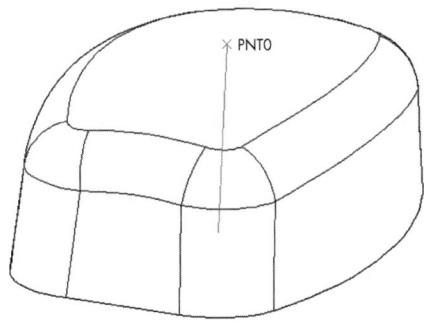

Datum point used to identify overall height of component.

Edges and Curves

Two functions under the Edges and Curves section are very helpful to the plastics designer. They are Max Dihedral and Short Edge. The dihedral angle is the angle found between two surfaces that intersect or join each other, as shown in the following illustration.

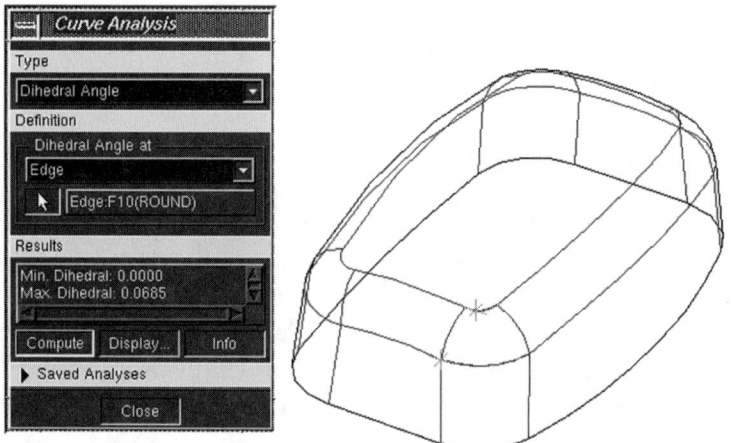

The Max Dihedral tool detects broken tangent conditions.

The Max Dihedral tool allows you to check the tangent condition of the edges of two surfaces and shows you where and what the maximum angle is between two surfaces that are supposed to be at an angle of 180 degrees to each other. The surface may have been created to follow the other surface exactly and at the same angle, but this does not always occur. You can go back and redefine the feature by adding more constraints or remodeling the surface feature. If you use this function as you design a part, you may be preventing future problems with the part when new features are added to the curve.

The Short Edge function is used to find edges that are small and not easily found. Perhaps a new feature was added and Pro/ENGINEER displayed a geometry check error stating that an item was too small. Using this function, you can make Pro/ENGINEER highlight the short section. You can see where the error occurred and then determine a method of solving the problem.

Minimum Radius

The Minimum Radius tool, found in the Surface Analysis section of the info tools, is useful in discovering why shell failures occur at a specific value. When a shell refuses to build, what is usually happening is that somewhere on the model's structure is a specific surface that will not offset by the desired value. Each surface patch on the model may be analyzed to discover the minimum radius of the particular surface. The results should point to the cause of the shell failure. Obviously, if the geometry cannot be repaired, a manual shell will have to be constructed. The following illustration shows the Minimum Radius tool applied to a surface patch.

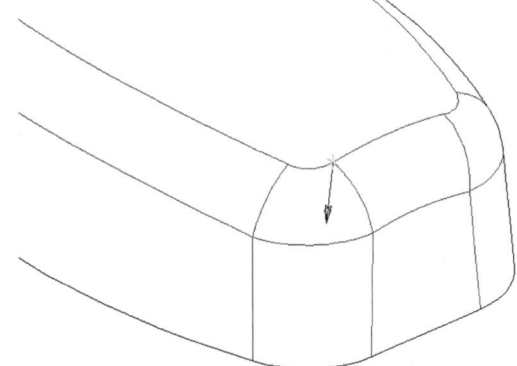

The Minimum Radius tool identifies the location of the minimum radius of a surface patch.

Thickness

The Thickness tool is used to determine the consistency of the wall thickness in your components. With this tool, you are able to set up parameters and use the software to slice the model through a series of predetermined slices or cross sections. The tool uses a color definition that references the maximum (displayed in the color red) and minimum (displayed in the color blue) thickness setting to provide you with the desired visual information. The following illustration depicts one of the cross-sectional slices through a component.

Thickness tool used to analyze for constant wall thickness.

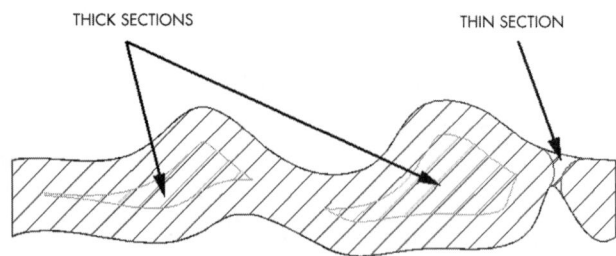

Draft Check

The Draft Check tool is commonly used in plastic design to verify the manufacturability of designs. The tool will identify surfaces and portions of your model that wander outside the anticipated design specifications. This tool should be considered a "must use" in plastic design.

Consider, for example, a component designed with a minimum draft angle on all surfaces of 1 degree, as shown in the following illustration. The most efficient way to use the Draft Check tool would be to set the desired angle to 0.99 degrees and select the actual parting plane to be the source for measuring the draft angle from. This would provide a situation where the software will color all of the surfaces greater than 0.99 degrees in one color and all of the surfaces less than –0.99 degrees in another color. The surfaces shown in the color range between 0.99 and –0.99 degrees would be considered problem areas.

Numerically Based Information Tools

Application of the Draft Check tool using 0.99 degree limits.

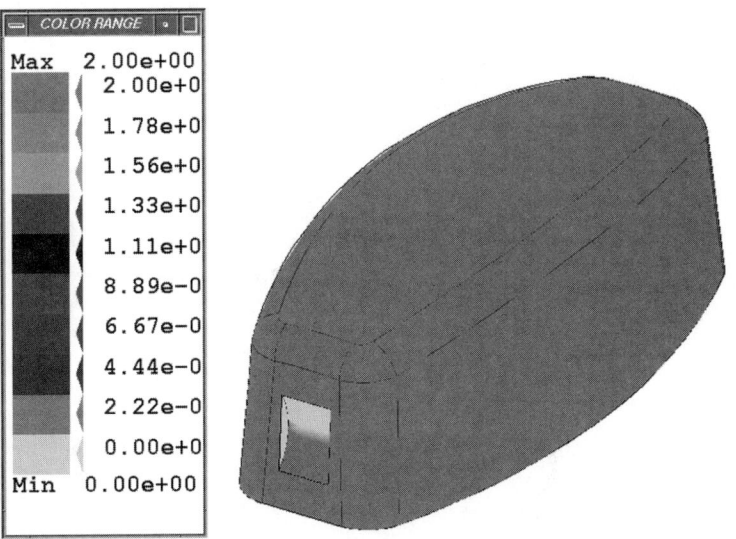

Therefore, if your component were displayed in two colors it would be perceived to be correct because the outer surfaces on the part would be one color and the inner surfaces the other. Where components are designed to have slides in the part, the Draft Check tool would need to be used, referencing all of the manufacturing parting planes.

In the following illustration, the walls of the component are drafted at 1 degree, with the exception of the intentional 1/2-degree undercut area on the end of the part. You will be able to identify this faulty area because the software will display it in a different color.

Undercut area to be checked using the Draft Check tool.

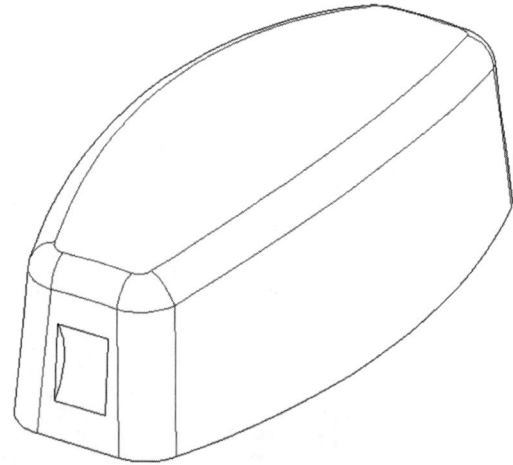

The numerical analysis tools constitute a suite of tools you can use to numerically test the integrity of models. Draft angles may also be checked manually using the Info Measure menu pick and selecting the Angle menu pick. It is also possible to verify the angle of edges and surfaces using this tool.

Visual Analysis Tools

Tools used to check curvature are typically visually based, and give you a good feel for the overall curvature changes across the structure of your design. The tool available for this type of analysis would be more often used to design high-curvature surface structures using Pro/SURFACE but may also be applied to generic Pro/ENGINEER modeling. Two of the tools available handle Gaussian curvature and slope analysis, both of which provide a color spectrum display.

You are also able to use the porcupine curvature tools to display curvature of surfaces and curve in your design. Porcupine tools are used primarily to identify bad curvature transitions or breaks in the tangent condition of surface boundaries within your model structure. Surface normal display can aid in troubleshooting imported geometry and is useful in visually aiding determination of the smooth flow of intersecting geometry.

Numerically Based Information Tools

The most commonly used tool in visual analysis is shading. Simply shading designs can show an incredible amount of information. This chapter explores techniques for using shading that may not be obvious to users in everyday use of the tool.

Gaussian Curvature

The Gaussian Analysis tool is found in the Surface Analysis section of the INFO menu. This tool's Gaussian curvature analysis display is shown in the following illustration. The display consists of a color spectrum comparing the smallest curvature with the largest one on the component. You are able to adjust the limits of the spectrum display and are able to spin the model with the color display active on your screen. This enables you to view all of the surface curvatures and transitions from various perspectives for best possible visual analysis.

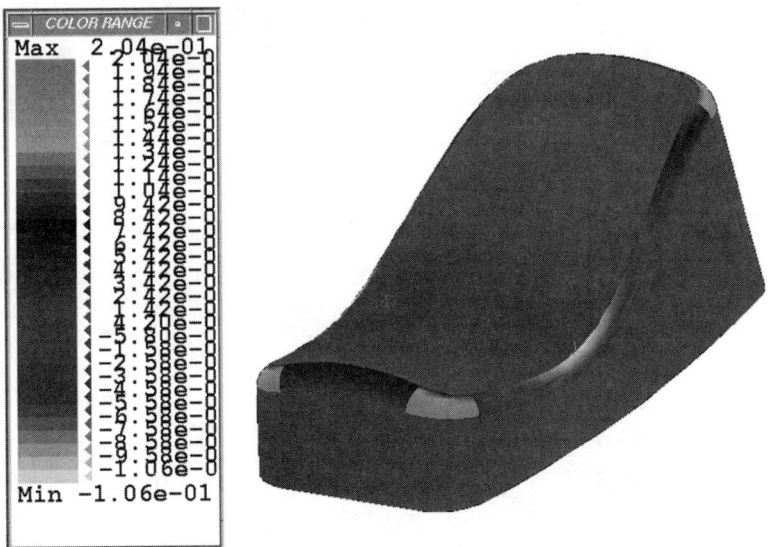

Gaussian Curvature Analysis display.

Slope Analysis

The Slope Analysis tool also provides a color spectrum display that references a datum plane and positive slope direction you select. The following illustration shows a component slope analysis display and the color spectrum. The tool helps you identify the relative smoothness of a model by highlighting abrupt changes in curvature in varied colors.

Slope Analysis tool.

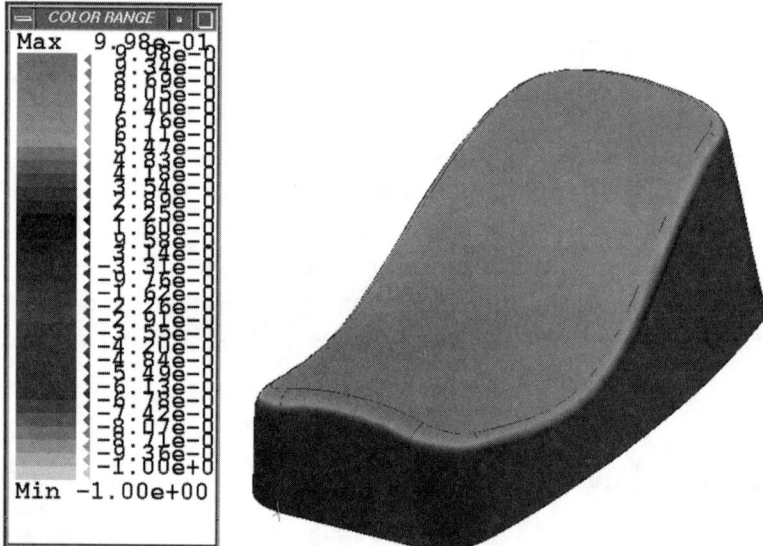

This tool can help identify problem areas in model quality and potential material flow problems. You can set the limits of the color display and spin the color display while the tool is active.

Porcupine Analysis Tool

The Porcupine Analysis tool may be used to evaluate surface or curve curvature. This is an exceptional tool for finding broken tangent conditions across surface patches. The tool provides a graphical display of lines normal to the surface structure or normal to the curve under evaluation. It is usually necessary to adjust the scale of the display to suit your viewing needs. Therefore, the scale of the display, the grid, and the density of surface normal display lines are adjustable, which accommodates viewing preference.

Numerically Based Information Tools

The Porcupine tool is commonly used by experienced users to evaluate their design at any time during the modeling cycle where it seems appropriate. The illustrations that follow indicate the flexibility of this tool as a design aid. In the following illustration, the Porcupine Analysis tool has been used on two adjacent surfaces in a model. The model chosen has a top surface of complex curvature made from joining surfaces. Note the position at which the curvature reverses on the surfaces.

Porcupine display of two adjacent surfaces.

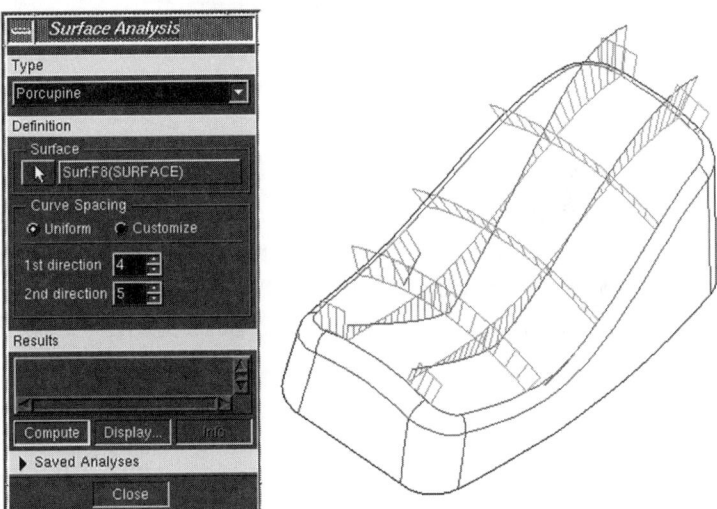

The following illustration is an analysis of two curves that mate tangent to each other. The curves have the same curvature at the mating point. Therefore, the height of the porcupine display is the same at that location. The curves shown are also tangent to each other (180 degrees).

Porcupine display showing two curves with the same curvature, tangent to each other.

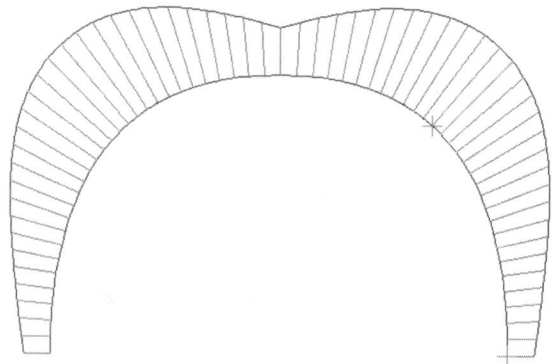

The following illustration shows the same two curves as the previous illustration but with the tangent condition intentionally broken at 174 degrees. Once again, the initial curvature is identical in each curve. Therefore, the height of the display is the same. Note the space between the two displays reflecting the 6-degree break in tangent condition.

Porcupine display showing the same two curves with a 6-degree break in the tangent condition.

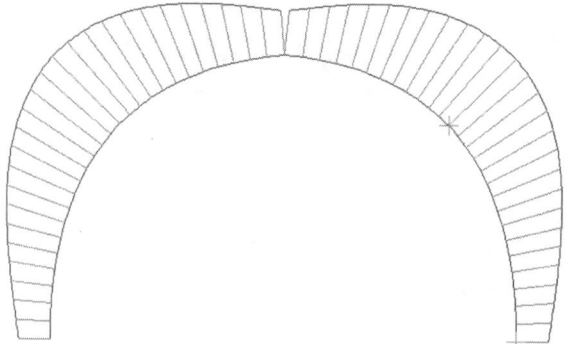

The following illustration shows two curves that are tangent but have a different curvature. Note the height difference in the porcupine display reflecting that at that point there is an abrupt curvature change.

Numerically Based Information Tools

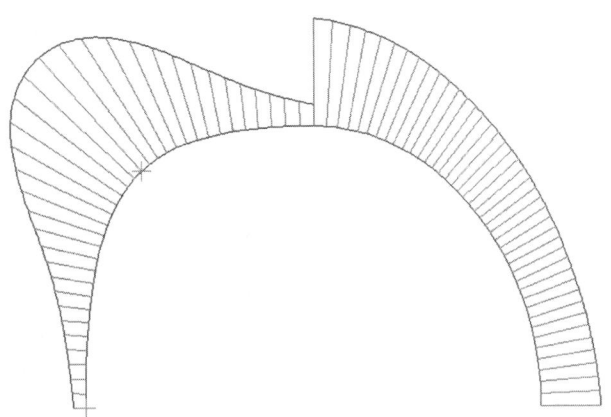

Porcupine display showing two tangent curves with different curvature.

Porcupine display can provide you with information about the curvature and tangent conditions of curves and surfaces. The tool is primarily a visual tool to help you identify relationships in adjacent surfaces and curves.

Surface Normal Vectors

The Surface Normal Vector tool is found in the SURFACE ANALYSIS menu in the Info area. This tool provides you with information about the shape and direction of surfaces. The tool may also be used to help evaluate imported geometry in the surface format to ensure that adjacent surfaces both have identical directions. The tool performs much like the Porcupine Display tool in that you have access to the scale and density of the normal arrow display. The following illustration shows an example of surface normal display.

Surface normal vector display.

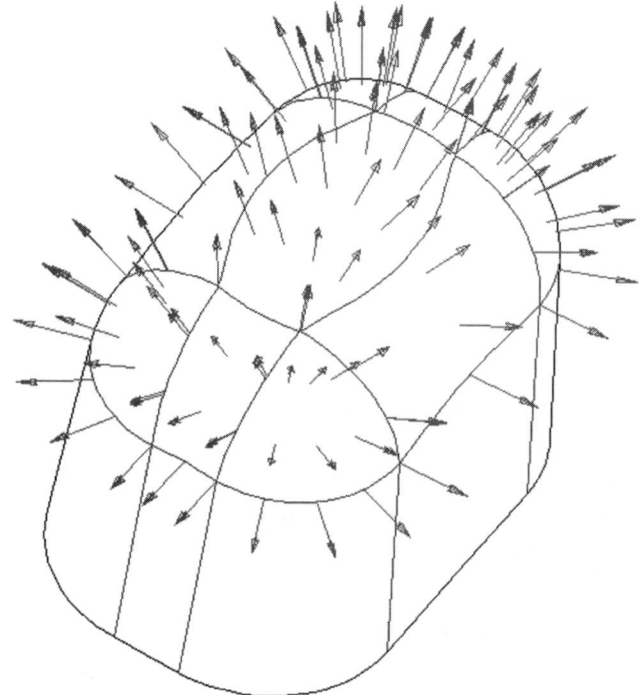

Shading

Shaded display of a design is one of the most useful tools for solid modeling using Pro/ENGINEER. These renderings provide a detailed look at the design in a form readily understandable. The important thing to remember about shaded images is that what you see is not necessarily what you need to settle for. The Cosmetic display menu allows you to customize colors and create more realistic renderings of models. The following list identifies some of the applications of shaded images.

- On individual components, surface patches may be colored independent of the default color.
- Entire part colors may be changed.
- You can define colors and store them in a color map for future use. The technique of using a common *color.map* file in an office environment

Numerically Based Information Tools

allows a design rendering to be displayed at more than one location in the same manner as displayed on the original terminal.

- Colors may be used to differentiate components in an assembly.
- User-defined colors may be made transparent to enable a shaded image of an assembly to have a transparent top housing.
- Display quality normal set at a value of 3 can be increased to a value of 10 for high-quality renderings.

This list only touches on the possible uses of the cosmetic appearance settings. As color maps can be user defined, it is a good idea to prepare one for use in your office. The text file may be located in a common directory and used by everyone accessing Pro/ENGINEER. The following example of a color map file contains only a few color settings. There are color maps in existence that employ over 100 colors.

Simple color map (color.map) file.

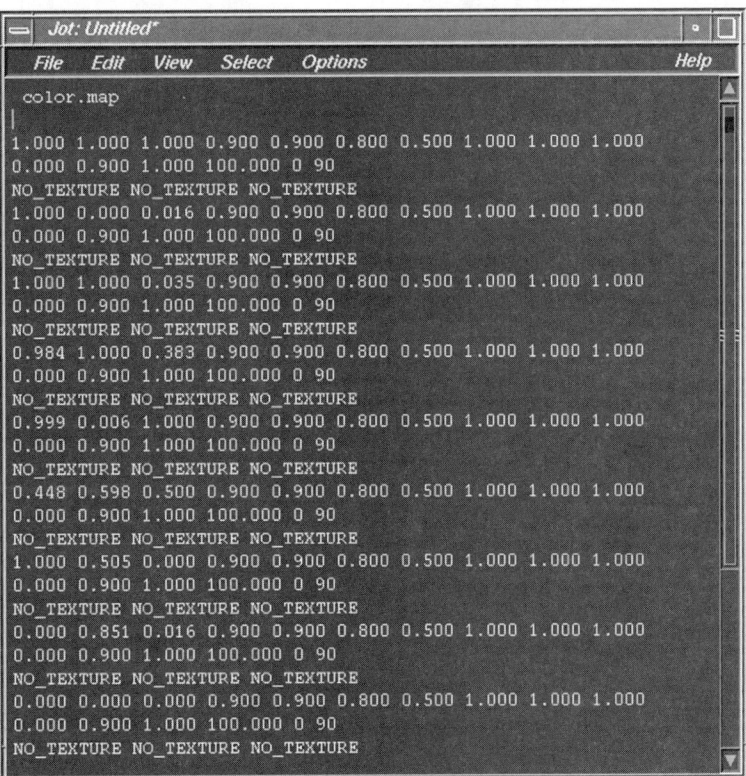

Display states are useful when looking at assemblies of components. Pro/ENGINEER allows users to view their assembly components in various states of display. In the following illustration, the top housing of an assembly has been assigned a display state of wireframe. When the shaded image is rendered, it is possible to view the internal components of the design through the solid top housing. This technique is useful when evaluating the fit of mating components or for presentation purposes.

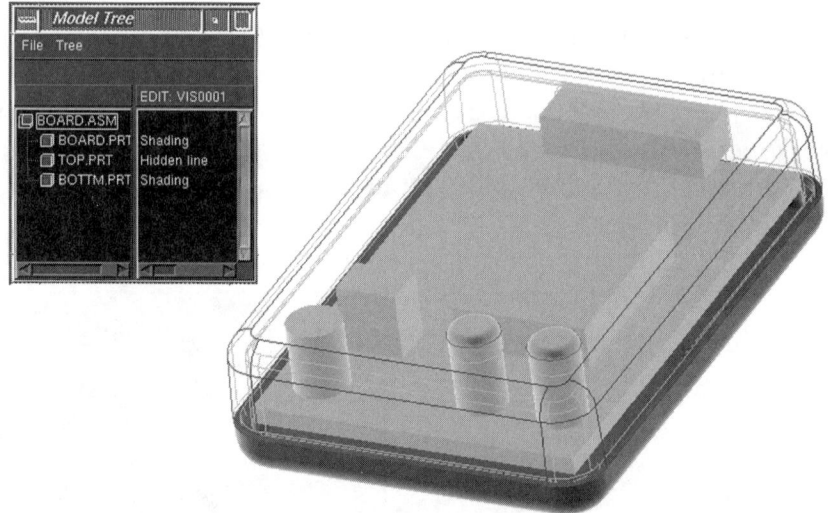

Shaded display of an assembly with a wireframe display of a top housing.

Clearance/Interference Checking of Assemblies

Pro/ENGINEER provides you with a number of techniques to check clearance and interference of components within a given design.

Clearance Checking

Tools are available in the Info section to evaluate clearance between surfaces. This tool is applicable to individual component surfaces as well as surfaces from different components used in a design assembly. The tool works much like the Distance tool and may be used to determine the minimum clearance between two surfaces. A common use for the tool is checking mating clearance of components.

Another tool for checking clearance is more visual in nature and is simply the cross-sectional display. By creating a cross-sectional display at a predefined planar position in the assembly, you are able to identify clearances within a model.

Interference Checking

There are three commonly used techniques to check for physical interference in an assembly of components. The down side of this process is that the design may contain some areas of intentional interference, such as crush ribs, press fits, snap engagements, preloaded levers, and ultrasonic welding energy directors. Designs that include intentional interference are more difficult to check for unintentional interference than those that do not purposefully include interference.

You can employ three techniques to solve for a problem area causing unintentional interference. Take, for example, a simple set of two housings with a plate internal to the housings. The plate has been modeled to intentionally interfere with the top housing on one side only.

The first technique is visual in nature in that a cross section will be defined and used to display the problem area. When creating cross section data for checking purposes, it is useful to model the assembly datum plane used for the planar cross section such that it is movable. This technique allows you to update the location of the cross section by changing the location of the datum plane.

When displaying the cross section, it is helpful on a complicated design to turn the display to no hidden lines. When the cross section is displayed, it is easier to identify the separate components. Note that cross-sectional display lines can be modified in density and angle to enhance identification of components. The following illustration shows a planar cross-sectional view identifying a problem area.

Cross-sectional view showing part interference.

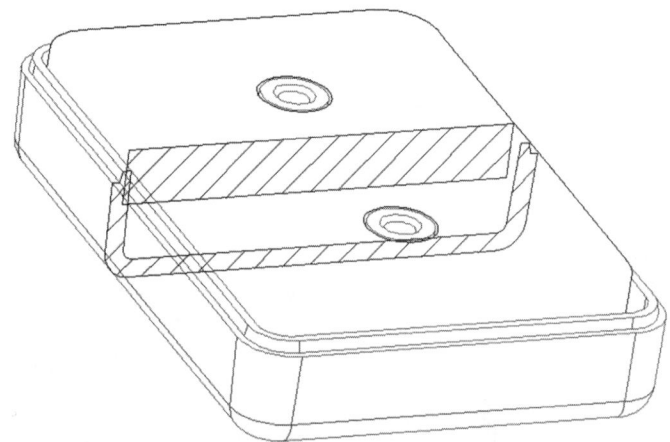

The second technique uses the Interference testing tool found in the INFO menu. When used in the assembly, the software will identify the position of the interference within the assembly. It is difficult to determine the extent of the interference and the number of interference locations using this technique. This tool is quite useful if the answer is that there is no interference. If interference is found, you may have to do further investigation to find the extent of the damage by zooming in on the area of interference and correcting your model. The following illustration shows the INFO Interference tool used to identify the problem area.

Numerically Based Information Tools 305

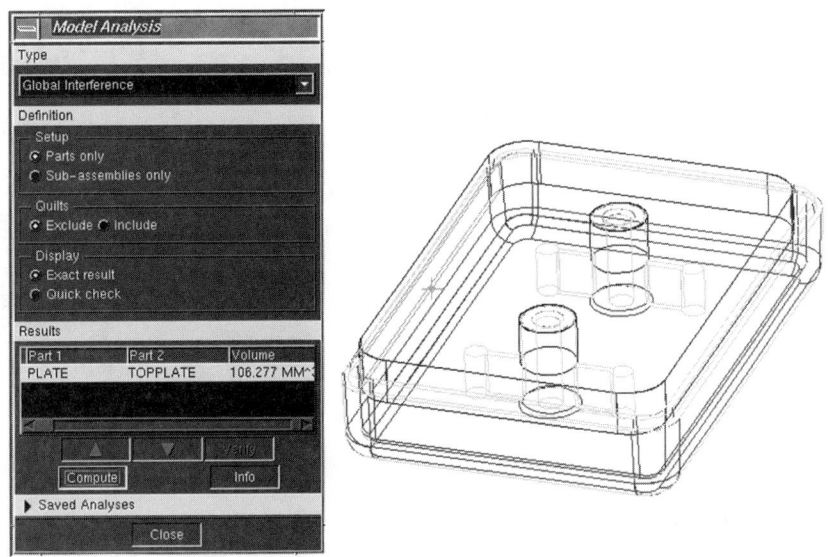

INFO Interference tool used to identify problem area.

The third technique enables you to not only find the interference but see for yourself the extent of the damage. This technique uses the problem part as a cutting tool to physically remove material from the housing. To keep things simple, take into consideration only the two parts of the assembly that do in fact interfere with each other. The following illustration depicts a shaded assembly showing only the top housing and the plate. Note that it is difficult to see the interference area of the parts.

Assembly of the top housing and plate showing the component interference.

Using the Merge Cut Out tool in the ASSEMBLY menu, the top housing is selected as the component to perform the cut to, with the plate then used as the cutting tool. The result, shown in the following illustration, clearly shows the area of material where the two components interfere with each other. The designer is also able to analyze the area to determine how best to solve the problem. If there were more than one location of interference, they all would have shown up as a result of employing this technique. This feature may be deleted from the top housing model once the investigation is complete.

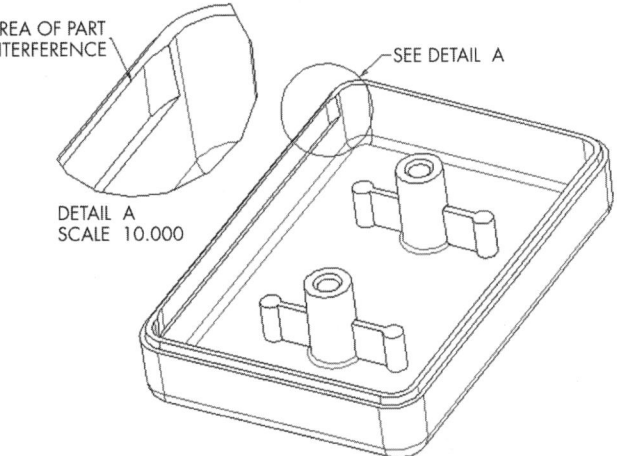

Top housing after Merge Cut Out applied.

Making Analysis Tool Use a Habit

It is important to get into the habit of testing your designs to ensure you are getting the desired results. Information analysis may be considered to be a burden to getting the modeling job done quickly, but in the long run it can save hours of rework time if problems are discovered early in the design cycle.

The most efficient strategy to ensure the integrity of your design is to test a model's problem areas frequently to ensure that draft features and so on are as they are intended. This concept is most useful when the root features of your design have been established and they are about to be used as parent features in the design.

Summary

The discussion in this chapter focused on analysis using the information available within the software. The chapter examined numerical analysis, visual analysis, and clearance/interference testing. Pro/ENGINEER provides designers with an excellent set of tools for evaluating finished and in-progress designs.

Chapter 18

Imported Geometry

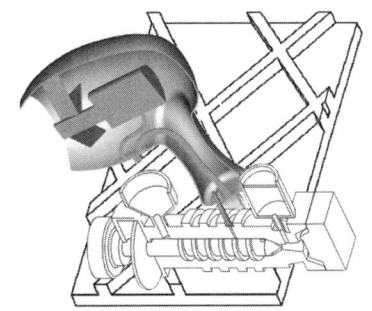

Geometry Reuse from Foreign CAD Systems

Introduction

Importing geometry into Pro/ENGINEER can be complicated. The purpose of this chapter is to demystify the world of imported geometry a little to better enable users to actually make use of imported geometry. The focus of the chapter will be on what to import, how to use it to your advantage, and what to avoid when importing geometry from foreign CAD systems. The STEP and IGES translation systems serve as the vehicle for discussing geometry imported into Pro/ENGINEER.

Some of the techniques for importing geometry and using it are covered to help you determine how to work with the imported geometry in Pro/ENGI-NEER. The Pro/ENGINEER redefine tools for imported geometry are covered to aid you in fixing or using imported geometry.

This chapter also addresses some of the common problems users are faced with when importing geometry from foreign CAD systems. The methods and associated data in the chapter provide a useful resource when you are importing geometry for your projects. The discussion also addresses what geometry to import, what you should build in Pro/ENGINEER instead of importing, and how to put the two together to model your project.

Geometry Source

Pro/ENGINEER reads foreign geometry in a number of formats. The formats used in this discussion are STEP and IGES. IGES is the older, more established procedure. STEP is a newer transfer technology.

Geometry is most commonly imported into Pro/ENGINEER by starting a new part and importing the STEP or IGES file into that part. The geometry is read into Pro/ENGINEER as a single feature. If there is a flaw or error in the geometry, the feature must be repaired if possible to create a useful piece of geometry. It is important to note that imported geometry follows the "garbage in, garbage out" philosophy. If the geometry has not been trimmed properly, or if you find the geometry has missing surface patches, you must fix the geometry prior to making a solid model from it. This can represent a good deal of work for a Pro/ENGINEER user when working on a complicated part.

For this reason, the creation of the foreign geometry on the host (originating) CAD system plays a critical role in the entire import procedure. If the geometry has been created at an inferior accuracy, you would need to make the Pro/ENGINEER model less and less accurate until you find the match that makes the geometry useful to you.

What You Can Import

The information here concerns importing model geometry, not the type of geometry found when translating 2D drawing information. The geometry is of a 3D nature. The format in which such geometry is transferred to the Pro/ENGINEER user plays a critical role in how it is used.

It is of importance to note that some geometry you might have to import may not have been created with the intent to go on to manufacturing. This is typical when importing design concept files made with the intention only for development of a product concept. This type of geometry is of little use to the designer embarking on a product design intended to be manufactured using plastic molding because the geometry lacks the features (such as draft) that make the design manufacturable. The list that follows provides an idea of what you may have to work with and how Pro/ENGINEER would react.

- *Point cloud data may be shipped to you for use in Pro/ENGINEER.* If this happens, the feature will read into Pro/ENGINEER as a single feature containing upwards of millions of points. This will be extremely difficult to use in Pro/ENGINEER unless you have the people who created the geometry at the host CAD system help you by identifying points that lie on a particular plane.

- *Datum curves may be delivered.* These work with Pro/ENGINEER provided the curves form proper patches for you to create your own surfaces on. The technique is highly dependent on the trim accuracy of the curves in the host CAD system.

- *Surface data, if any, can come with a variety of characteristics.* The surface structure may form a closed surface patch boundary or may have open places. This condition could exist using surfaces imported using IGES or STEP. If the surface structure is open, you must close it in Pro/ENGINEER to obtain a solid from the imported geometry. Not only must the designer close the surface but do it in a manner that suits Pro/ENGINEER. The surface patch structure of the model must form a closed boundary. This is sometimes referred to a watertight surface skin. Once a watertight skin has been established, the geometry may be filled with solid material.

- *Solid geometry may be imported into Pro/ENGINEER using STEP as the translation tool.* Files prepared to transfer in this form may have been made as solid or watertight structures by the host CAD system. Pro/ENGINEER will fill a watertight closed surface structure upon import. Alternatively, you can redefine the surface structure so that it can be filled.

> ✓ **TIP:** *If at all possible, and early in the design cycle, work with the person supplying the foreign geometry. You will obtain the best solution to your project requirements and save time in the long run.*

Defining Import Requirements

Often users are faced with trying to import a geometric model that describes a part or product in its entirety. This can become quite frustrating when dealing with tiny surface patches or trying to repair fillet radii that are easily created using Pro/ENGINEER. An example of this would be importing a complete model of a component when the only complicated thing about the imported geometry is the top surface. It would be most efficient in this case to import the top surface only, and use Pro/ENGINEER to do the rest.

This concept sometimes results in users rebuilding geometry in Pro/ENGI-NEER, using the imported geometry as a guide. It would be better if at the outset the user were able to model uncomplicated geometry in Pro/ENGI-NEER and import just the top surface. The following illustration shows an example of this type of model. There is a complicated top surface, but the shape, side walls, and fillet radii would be quite simple to produce in Pro/ENGINEER.

Import geometry with a complicated top surface.

The following list of characteristics highlights how this project would be better handled by importing the top surface only. If for any reason the top surface changes during the model design phase, a new surface can easily be substituted for the existing one using the redefine tools. This process maximizes the flexibility of Pro/EGNINEER modeling if there are product changes to consider. The imported geometry is not modifiable, only replaceable.

- There is a complicated top surface that should be imported. An alternative to this would be to import the curves from which the surface was created and construct the surface in Pro/ENGINEER.

- The product is basically rectangular, with corner fillets. This geometry is easily modeled in Pro/ENGINEER.
- There is a fillet round all around the part at its top. This geometry is also easily modeled in Pro/ENGINEER.
- Assuming there is a constant wall thickness, this design would be very easy to model in Pro/ENGINEER as a shell feature.

Examining these characteristics of the geometry to be imported indicates that the bulk of the geometry would be well suited to creation in Pro/ENGINEER, which would provide the fringe benefit of being modifiable after creation.

Features Difficult to Import

Features most difficult to import into Pro/ENGINEER are usually those that represent cosmetic value in a component. Rounds, grooves, slight recesses, and raised areas provide the most difficulty because they are usually made with tiny surface patches. This type of geometry is also reasonably simple to create from scratch using Pro/ENGINEER once a solid is realized. Therefore, it can often be counterproductive to import such features.

Surfaces untrimmed or trimmed to a poor accuracy are also difficult to import because these tend to contain small gaps or overlaps in the surface structure that cause problems with the import tool.

> ✓ **TIP:** *For the best results, try to import only the geometry that is difficult to create in Pro/ENGINEER.*

If the host CAD system has produced an export file with poor accuracy, Pro/ENGINEER will have difficulty working with that file. If you are required to import geometry from an inaccurate modeling system, leave the fillet radii and small features behind if you can. When sharing data with another CAD system, settings are available that can help the CAD systems communicate. CAD software vendors' help lines can aid you in determining how to set up your configuration files to share data with foreign CAD systems.

Redefine Tools for Imported Geometry

It is often necessary to fix geometry that has been imported from a foreign CAD system to repair surface patches and vertex locations prior to obtaining a solid model from the Pro/ENGINEER geometry. When redefining imported geometry, tools are available that aid you in correcting small problems in the imported geometry file.

> **NOTE:** *Some problems may not be solvable and you may have to resort to reimporting another model or trying to build another model in Pro/ENGINEER using the imported geometry as a guide.*

The following is a list of troubleshooting options to consider when faced with a problematic import file using the STEP or IGES formats. These items may help determine why a model exported as a solid or closed surface model fails to read into Pro/ENGINEER as a solid model.

- View the model being imported, looking for holes or gaps in the surface structure. If there are surface patches missing they may be modeled using Pro/SURFACE to complete the watertight surface structure, with a solid obtained after this process is performed.

- View the model, looking for open boundaries in the surface structure. The surfaces must be joined in Pro/ENGINEER prior to obtaining a solid from the geometry. Joined surface boundaries are shown in magenta and open surface boundaries are displayed in yellow. By looking at the yellow boundaries, you are able to determine how difficult the problem would be to correct. You may be able to correct the problem simply by lowering the default accuracy of your part to close the open boundaries.

- Using the feature redefine tool, you may be able to correct open surface boundaries by zipping the gaps in the surface structure. When tackling this job, first unjoin all of the surfaces prior to zipping the gaps. After the gaps have been zipped, you may rejoin the surface structure. If open boundaries still exist, the process may be repeated using a larger gap setting. If this still fails to fill the gaps, the faulty boundaries may be corrected individually using the Fix Boundary pick in the menu.

- You are also able to repair faulty vertex location, edges, and tangent conditions using the Redefine tool.

Using these techniques, you should be able to import any reasonably accurate geometry into Pro/ENGINEER and create a solid model from it. Note that it is sometimes best to import a wireframe model of geometry instead of carrying the overhead of a complex surface file in your part. If geometry is imported for reference only, try to make sure that the imported feature is not referenced when you create your model. This allows you to delete the imported geometry feature once the project has been finalized.

> **NOTE:** *Be aware that imported geometry is not easily modified. If this is unacceptable, you have the alternative of using the imported geometry to create your own model.*

IGES and STEP

IGES – Initial Graphics Exchange Specification

IGES geometry transfer has been around for many years and is a standard for exporting geometry to foreign CAD systems. IGES allows for the transfer of specific types of geometry and is suitable for transferring points, curves, and surfaces in a very accurate manner.

STEP – Standard for the Exchange of Product Model Data

STEP is relatively new and has the potential to become the global standard for transferring geometry. Using STEP as the transfer tool, your imported geometry can take on characteristics not available in IGES. Some CAD systems allow the export of solid geometry using STEP.

One advantage of using STEP as the transfer mechanism is that STEP appears to provide additional intelligence to models imported into Pro/ENGINEER.

That is, there is at least a partial transfer of design intent from the host CAD system to Pro/ENGINEER. For example, surfaces made tangent in the host CAD system will be recognized as tangent surfaces by Pro/ENGINEER. It would not be inappropriate to state that fillet radii may be imported into Pro/ENGINEER using STEP as the transfer tool.

Import Geometry Project

The import geometry feature selected for this project could have been created on a number of foreign CAD systems. The assumption is made in advance that the designer on the host CAD system has intentionally made the component manufacturable and has added the appropriate draft features and manufacturing data.

The product chosen for this project is an industrial design file exported using the STEP format. However, IGES would work as well if the surface structure were trimmed to an appropriate accuracy and joined. The following sections take you through the phases of the project.

EXERCISE

Phase 1: Set Up the Project

Start a new part in Pro/ENGINEER to import the foreign geometry feature into. Use an accuracy of .0008 to start with. This number may be tweaked at a later time if necessary. It is common procedure to use a start part program that will set up accuracy, view names, and default datums for you. The coordinate system is also included in the start part and may be used to locate the imported feature. If the coordinate system does not suit your orientation needs, create one that does.

Phase 2: Import the STEP Geometry

Create an import geometry feature using the coordinate system established in the start part. While importing the feature, Pro/ENGINEER will keep track of the entries in the imported geometry file in a log file. If something goes wrong during the import, you are able to use the information in that file to help you troubleshoot the problem.

If the import feature is missing surface patches or has boundaries that are not joined, the part will read into Pro/ENGINEER as a surface model. You will be able to identify the problem areas in the file as open boundaries by the color in which they are displayed. Yellow boundaries signal a problem in the import geometry of an intended watertight surface model or a set of merged surface patches (quilt).

1. Using the Interface tool, import the STEP file into the part.
2. If problems exist, view the model for open areas or boundaries shown in yellow.
3. If repairs are necessary, complete them at this time.

> ✓ **TIP:** *Shading the model will show holes in the surface structure. You may also color the external side of the surface to help identify a problem. This is accomplished using the Cosmetic Appearance tool.*

The following illustration shows problems in the surface data: open boundaries where the small fillet radii exist on the geometry.

Open boundaries found.

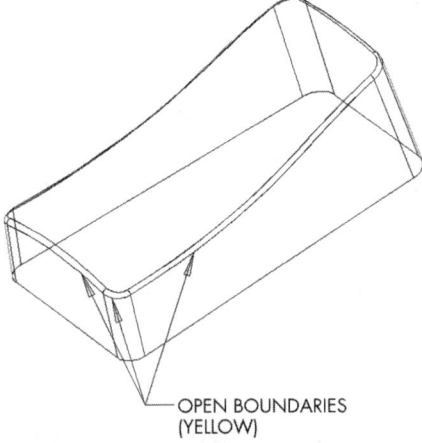

OPEN BOUNDARIES (YELLOW)

Even after zipping the gaps, you may not be able to maintain reasonable part accuracy and read the geometry into Pro/ENGINEER without experiencing geometry problems.

> ✓ **TIP:** *You might have better success using STEP than IGES. Most of the popular CAD systems available have added STEP to their suite of export tools. STEP is also used to export solid geometry, making it a good match for Pro/ENGINEER. Try experimenting with STEP and IGES and make your own decision.*

You may use the redefine tools to close the gaps in the file, or you may stop and have a look at what geometry you are importing and how you may be able to simplify the import process.

Phase 3: Proceed or Retreat

Features such as fillet radii are difficult to import accurately due to the program's stringent boundary conditions. These features represent the lion's share of problems with importing geometry into Pro/ENGINEER. It is ironic that fillet radii are one of the easiest features to create in Pro/ENGINEER.

When deciding on whether to proceed or retreat, it should be mentioned that the Pro/ENGINEER tools for repairing imported geometry have become very sophisticated in their capabilities in the later releases of the software. This does not mean, however, that the user should endeavor to fix any model regardless of how bad it is just because the tools are provided. Quite often, looking at the import task in a different way will simplify the product modeling task and allow much more design freedom to the project team.

If you were able to request another copy of the import feature without the finishing radii, the geometry would be much easier for the software to calculate when importing. This would be a win-win situation, as you would end up with a more flexible model regenerated much faster in Pro/ENGINEER. The radii would be modeled in Pro/ENGINEER and would be modifiable if changes were required. The following steps would allow you to re-import the feature into your design. The illustration that follows shows a new STEP file imported without fillet radii.

1. Delete the original import feature from your part file. You may also redefine the feature if you wish and select another file to import.

2. Read in the new STEP file. If it does not come in as a solid, redefine the feature to ensure that the boundaries are closed and the surfaces joined.

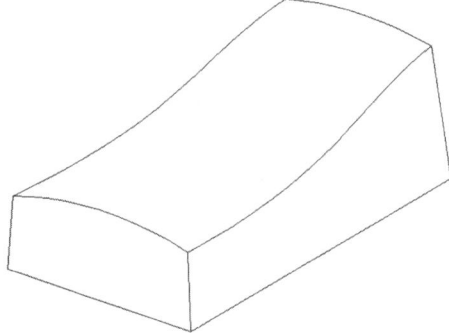

New STEP file imported without fillet radii.

Phase 4: Finish the External Geometry

Once the imported geometry has been successfully read into Pro/ENGINEER and made solid, you are able to quickly add radii to the external geometry to finish the external shape. This model, shown in the following illustration, may be used to proceed with the detail design.

Radii added in Pro/ENGINEER to complete the external shape.

This completes this project on importing geometry using STEP. The detail design may now proceed, wherein the shape is divided into manufacturable piece parts.

IGES and STEP

LOG Files

The log file shown in the following illustration shows the entities read into Pro/ENGINEER during the import process. If you have difficulty reading in a file, you are able to consult the log file, which is a record of the import process. This is useful when troubleshooting problem geometry.

Partial STEP log file.

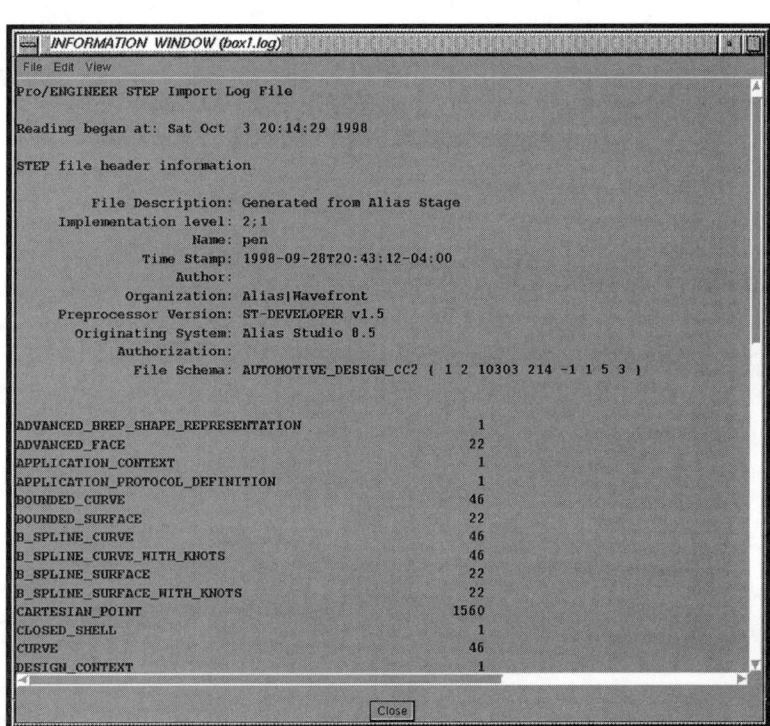

Dealing with Change in Imported Geometry

Users often feel helpless confronted with shape changes to Pro/ENGINEER models for which imported geometry has been used because the product or component was based on a single import feature. When many subsequent features have been added to the original import feature, this situation can be very difficult for a designer trying to work to an aggressive deadline. The parent/child relationship developed within the Pro/ENGINEER model tree structure sometimes leads to a misunderstanding that the modeling task must be started over from scratch should a change to the geometry be required. However, this is not the case.

Pro/ENGINEER provides tools for re-importing geometry and gives you a chance to save the parent/child relationships previously defined by reallocating edge and surface references from the previous feature to the new import feature. It is important never to start over until you have tried redefining the import and adding the new shape to the Pro/ENGINEER model tree. This is much easier if the import feature is a single surface rather than a complete product, but the re-import tool will work in either case.

Re-import Example

The housing shown in the following illustration has been created from foreign geometry and the design has progressed to add the shell and some of the internal features. Assume that you have taken care to create as many features as possible using the Thru Next menu pick. This enables the features in your design to rebuild easily if the external reference surfaces are replaced somewhere in the modeling cycle.

Import feature with subsequent features added.

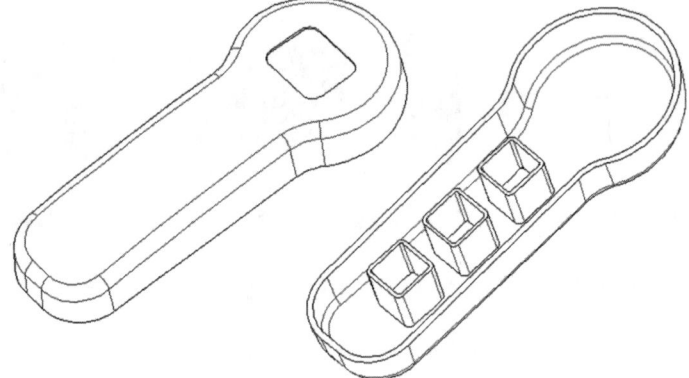

The shape of the new file is similar to the existing geometry. However, its top surface has a different shape. The file shown in the following illustration will replace the original import geometry feature in the part file.

File showing the new import feature.

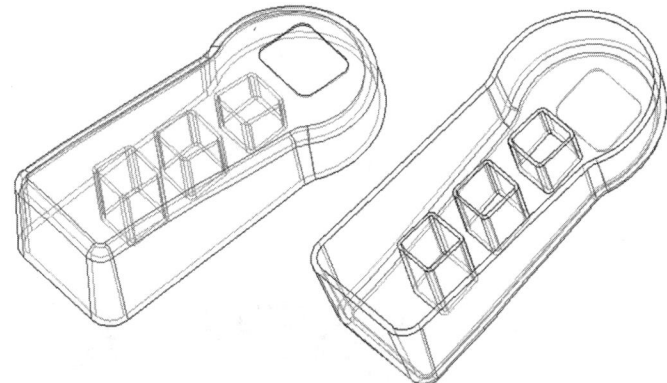

Using the Redefine menu pick and selecting the submenu FROM FILE under the REDEFINE menu will enable you to replace the old import file with the new one. Once this has been done, the new shape will appear in the part window and a second window will open with the old reference part in it.

The software will then start to quiz you to select the appropriate edges and surfaces on the new part that coincide with references previously found in the original import feature. The following illustration shows how the screen will look during the file substitution phase. The new import feature is in the main part window and the software has opened a second window with the old import feature referenced.

Re-import process showing the Pro/ENGINEER screen.

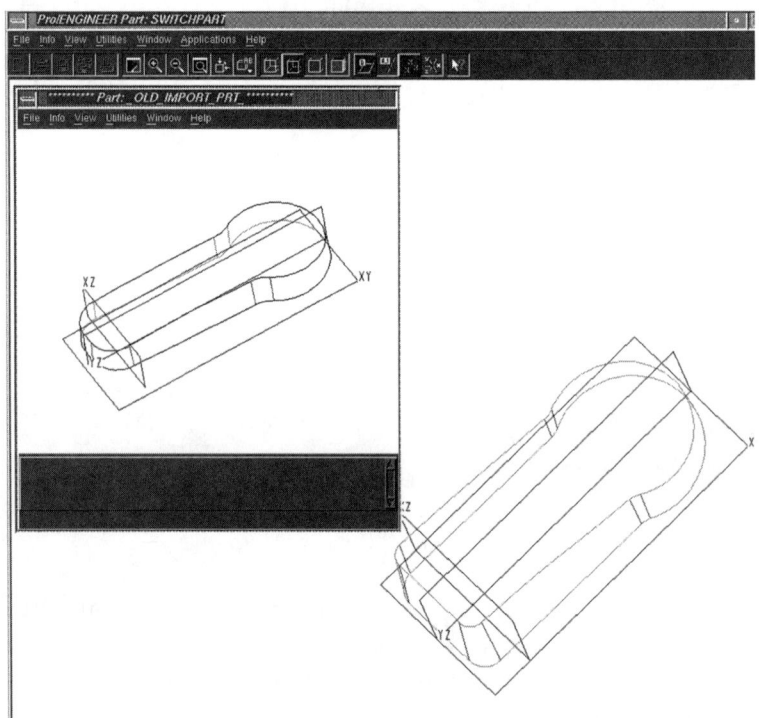

Once the software is satisfied that you have indeed replaced the old feature's references, the second window will disappear and the regeneration of the rest of the features will commence. Any features that fail upon regeneration should be repairable if there are appropriate references to which to reassign children.

> **NOTE:** *An import file does not have to be replaced with the same file type.*

If you originally used an IGES file in your design you may now replace it with a STEP file if you wish. There is quite a degree of freedom in the redefine tool in Pro/ENGINEER. If you desire even more freedom, you may consider the approach to importing geometry described in the section that follows.

Logical Import Process

Import geometry does not have to be totally non-modifiable. The manner in which you import the design is definable. You are able to assume at least part of the control. If you were to break the job up into smaller pieces you would not only make the calculation easier for the software but give yourself a greater degree of freedom.

Consider the following example of a holding base for some product. Everyone on the product team is reasonably comfortable with the overall shape of the base housing but are still working on a number of concepts for the product. This situation is holding your product design from proceeding, as you would like to use the file with the pocket already in it for the product.

If the job were split up into more than one import file, you would be able to proceed with the product detail design and substitute the correct pocket in at a later date. To do this you must consider importing a file for the base housing and a second file for cutting out the product pocket. You may be able to expand this concept in actual use to the point where you would have one import file for the basic product shape and a number of import files representing things that need to be cut out of or added to the original file. This type of modeling is much like using building blocks. The file for the original shape is shown in the following illustration.

Base housing import feature representing the product basic shape.

Once the basic shape is read into Pro/ENGINEER, it is made into solid geometry. Any subsequent styling features such as rounds may be added at this point if they do not affect the pocket area of the charging base. Internal design features should be left until the product shape is subtracted from the base housing. The surface structure for the pocket is shown in the following illustration.

Import feature representing the pocket

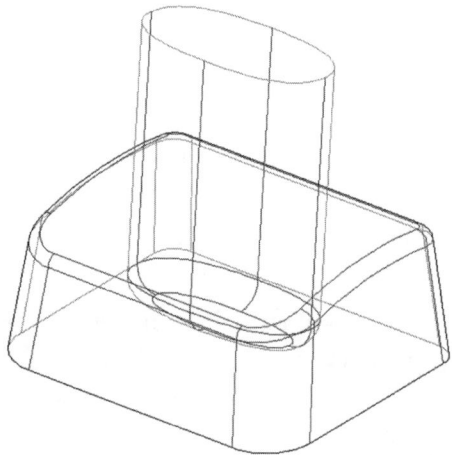

The surface model of the product pocket is added to the original part file and used as a cut to obtain the finished external design. If the product shape representing the pocket changes, you are able to re-import a substitute file reflecting the latest design. The product with the pocket cut is shown in the following illustration.

Pocket is cut from the original import solid geometry.

This technique may add a huge degree of freedom when using imported geometry to create Pro/ENGINEER components or product designs.

Summary

This chapter has covered various techniques for importing foreign geometry into Pro/ENGINEER. The focus of the discussion was on the import data structures IGES and STEP. The basic information presented is applicable to imported geometry from any import option supported by the Pro/ENGINEER interface tool.

Discussion of what to import and what not to focused on determining the best possible means of creating the geometry for a design. The import geometry feature was covered in detail, as well as the capabilities of the Redefine tool. Tips on importing geometry included a discussion on breaking the import task down into smaller, more manageable features.

Index

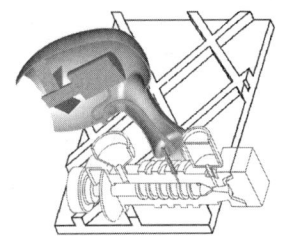

Numerics

3D modeling
   critical specifications, identification of 92

A

ABS
   See acrylinitile butadiene styrene (ABS)
accumulator, foamed parts 23
acrylinitile butadiene styrene (ABS) 52
Advanced round feature 129
Advanced Utilities pick 200
Advanced Variable Section Sweep protrusion,
   ultrasonic welding 104
alignment ribs 238
analysis of design
   See design philosophy
analysis tool use 307
assembly jig, ultrasonic welding 222
ASSEMBLY menu 200
assembly requirements for end users, considerations
   of 57
Auto Blend option 127

B

base design, creating 208
battery location, start model 179
bend table 280
bending and deflection guidelines 139
blends, neck section of bottle 258
blow mold designs 257–264
blow molding
   cavities, molding process 28
   designing a bottle for 258–263
   parting lines 29
   three stages of 26
   tube formation 27

uses for 257
bottom-up design approach 64
boundaries, open, IGES and STEP translation
   systems 318
Boundary Loop option 209
breakage prevention, energy absorption and 46
Brinnell scale of hardness 280
buried features
   design planning and organization of 83
   tips for identifying 98
butt weld joint method 223
button designs for complex parts, versatility of 41
button labeling for part design, functioning of 50

C

CAD file, use in plastic molds 32
CAD systems, foreign
   See geometry, imported
CAE
   See computer-aided engineering (CAE)
cavities
   blow molding 28
   plastic molding 5
cellular bubbles in foamed parts 24
center of gravity 281
ceramics, electrical insulation properties of 36
chemical blowing agents for foamed parts 24
clamping and snap fit joints 111
clamping features 193
clamping force, plastic molding requirements for 19
clean sheet approach
   See top-down design
clearance and interference checking of
   assemblies 302
clearance and interference tools 302–306
color-coded displays, interference tools and 287
component accrual and assessment 64

component design
    See thin wall component design
component fastening 101–112, 217–236
component replacements with one plastic part 203–215
component repositioning 63
    initial component layout assessment 63
component shape and overall part strength 135–142
component-level planning
    See design planning and organization
components
    gathering and assembling known 65
    modeling 200
    off-the-shelf
    positive matching of surface structure and shape 73
    splitting models into 55
    tooling costs, avoidance of 65
computer-aided engineering (CAE) 280
computer numerically controlled (CNC) milling 30
computerized process control, technology advancements 238
concave surfaces and durable surface finish 49
condition 280
convex surfaces and durable surface finish 49
cooling lines 20
    speeding of cooling process 13
Copy Geom option 201
cord grommet housing design, aids in 75
core, plastic molding 5
cosmetic appearance 39–41
Cosmetic Appearance tool 317
cosmetic finish requirements for end users, considerations of 57
cosmetic shape of product, identifiable restrictions 60
costs, associated 50
criteria for material selection 51
critical dimension drawing, items to include 94
cross section mass properties 283
crowning 130
    plastic surface, large 181
crush ribs 219
curves
    See edges and curves
cut profile, sketch of 241
cut trajectory basic shape, creating 241
cycle-time reduction for injection molding 21

D

datum curve 75
    cut created following 181
datum curve menus, labeling 262
datum curves
    imported model geometry 311
    product design modeling techniques 74
    root features 89
    skeleton model technique 201
datum plane, Through/Axis 144
datum points
    product design modeling techniques 75
    using 289
Datum section 164
datums, creating new component containing 207
deflection and material strength 52
deflection guidelines
    See bending and deflection guidelines
deflection resistance 48
design considerations
    product features 64
    See also material property and design considerations, other
design decision factors, associated costs 50
design limitations
    See specifications, known
design philosophy 238
    and master part plan 198
    applying 171–172
    initial analysis and project planning 55–81
design planning 205–215
    drop test requirements 58
design planning and organization, component-level 83–112
design requirements, identifying product features for 61
design splitting
    front and back piece parts 70
    left and right piece parts 71
    top and bottom components 69
die for extrusion molding 25
dielectric constant of material for electrical insulation properties 39
Dim Bound feature 160
dimension drawing, critical
    specifications of 93
dimension relations 161–163
dimensions, critical product, proper functioning of 93

Index

dimensions, mating features 94
disassembly restriction requirements for end users, considerations of 57
Distance tool 302
door finishing features, adding 190
door stop feature 192
draft 7, 120–123
   cosmetic features 115
   split 122
   See also ribs and draft
draft and cosmetic features, adding 182
draft angles 7, 8
   Info Measure option 294
Draft Check tool 292
draft
   adding 230, 242
   external for product design additions 183
drop test requirements 58
   end users, considerations of 57
duplicates, creating 210
durable surface finish 49

E

ears 131
edge mark, slides and 12
edges and curves 290
Edge-Surf feature 126
EDM (electrical discharge machining) 14–15
   electrodes 14
   milling machines 14
   vents 15
ejector pins
   flash lines 10
   material flow 156
   removing plastic part from core 9
electric razor wrap-up 72
electrical discharge machining
   See EDM
electrical insulation properties 36–39
electrically powered equipment
   electrical insulation properties for 38
electrodes, EDM process 14
electronics assembly, modeling plan 206
electronics mounts, placing 213
emissivity 279
end users, considerations of 56
energy absorption 45
   breakage prevention 46
   entropy 46

feature, illustrated 47
heat energy 46
interlocking features 47
rounds 47
sound energy 46
stress concentration 47
energy conduction
   See heat factors
energy director ultrasonic welding of parts 104
energy director method 225
energy, air drag effects of energy absorption 46
energy, potential and energy absorption 46
entropy and energy absorption 46
epoxy molds, rapid prototyping and 31
etching, cosmetic appearance 40
exothermic reactions 32
external rounds, adding 242
extrusion molding 24
   die use in place of mold 25

F

fabrication
   See plastic molding and fabrication
fabrication methods
   hoppers, purpose of 17
   polymers, air-drying machines 17
fastening
   See component fastening
features
   difficult to import 313
   evaluation of 163
   finishing 233
   hold-down 235
   patterning 231
   rename 89
   snap fit 10
   unattached 73
fillers
   adding 53
   defined 51
   glass fibers 51
finite element analysis 280
finite element tool 285
fit and form
   importance to end users 56
   product's outer shape, laying out 66
fit, form, and function of component
   specifications, identification of critical 93
Fix Boundary option 315

flame-retardant requirements for end users,
      considerations of 57
flame-retardant tests 58
flash
   defined 158
   managing 158
flash and material flow 156
flash lines and ejector pins 10
flexible material for live hinges 43
flip-up door, creating 187
flow length 152
flow runner for material flow 155
foamed parts
   accumulator 23
   cellular bubbles 24
   chemical blowing agents 24
   injection molding 23
   viscous plastic 23
footpad, patterning of 210
form and fit of final product,
   rapid prototyping for 30
friction shock, electrical insulation properties 37

G

"garbage in, garbage out" philosophy 310
gas assisted injection
   See injection molding
gates 159
   cost considerations 20
   plastic molding 12
   See also sprues and runners
gating, sequential,
   technology advancements 238
Gaussian curvature analysis tool 294–295
gears, interference of
   maximum/minimum tolerance modeling 161
generic start part 176
geometric datums 94
geometric tolerancing 94
geometry
   finishing external 320
   imported 310
   logical feature creation process 96
   reuse from foreign CAD systems 309–327
   shelling 211
   source 310
Graph function 163–164
grooves, features difficult to import 313
Group function 195

Group option 231
gyration, radii of 282

H

hardness 280
heat dissipation problems, solving 135
heat energy absorption 46
heat factors, energy conduction 36
heat staking 218
heat staking technologies for ultrasonic welding 102
Helical sweep function 260
high profile 139
history-based modeling system 87
hold-down features 149, 235
hold-down locations for component design 172
hold-down posts, adding 214
hold-down ribs 178, 192–194, 199, 215
hollow
   See cavity
hoppers, purpose of 17
horn for ultrasonic welding of parts 102, 103
housing and pivot lever,
   snap fit joints for 107

I

IGES files, exporting 93
IGES translation systems 309, 310
      imported geometry, redefining tools used for 314
import geometry project 316
import requirements, defining 312–313
imported geometry
   changes in 321
   redefining tools used for 314
industrial design departments, considerations of end
      users 56
inertia
   moment of 281
   principle moments of 282
Info Draft Check tool 247
INFO Interference tool 304
Info Mass Props option 272
Info Measure option 294
Info Measure selection 156
Info Measure tool 289
INFO menu 295, 304
Info section 263, 302
information and clearance/interference for multiple
      component design 287
information tools 287–302

Index

initial bend y factor 280
initial component layout assessment 63
   piece parts, individual 63
   rough-modeling approach 62
   shape dimensions, modifying 64
   skeleton models 63
initial graphics exchange specification (IGES) 315–327
injection molding
   cost considerations 20
   cycle-time reduction 22
   foamed parts 23
   gas assisted 21
   mold flow analysis 23
   process 16
   shot size 17
   sink marks 21
   traditional and thin wall 16
injection molding product design exercise 169
injectors 17
inner chassis, clamping 192
inserted areas 159
intelligent start part plan 175–197
interference checking 303
   component fastening 217
   finished butt joint feature 224
   interference tools 302–307
interlock detail, creating 185
interlock geometry, creating 244
interlock lip 197
   securing 193
interlock ribs 244
interlocking features and energy absorption 47
interlocking lip
   logical feature creation process 97
   ultrasonic welding of parts 104
interlocking rib 149
internal ribs, adding 245
Intersect Merge feature 274

K

keys and jack locations 180
   master part features, desirable 199

L

labeling
   bottle 261
   buried features 99

styling features 175
   See also datum curve menus
laminated object modeling
   See paper
lasers for rapid prototyping parts 30
latch modeling process 266
layout assessment, initial component
   visualizing product designs 62
LCD screen modeling plan and process 239
left-hand model, adding to assembly 267
lens window logical feature creation process 97
line of draw 170
lip feature 132, 224
lip selection process 187
live hinge, flexible material for 43
location points for skeleton model technique 201
LOG files 321
logical feature creation process 95–101
   housing example 96
logical import process 325
low cross section 139

M

managing tolerances and relations 160–163
manufacturability, designing for 151–165
manufacture cycles for molds 4
manufacturing technology limitations 58
marketability, cosmetic appearance and 39
mass density 278
mass moment of inertia 281
mass properties
   applying 277
   cross section 283
mass properties file 281–284
mass property function 263
master model technique 171
master part approach to modeling CD player 197
master part features
   desirable 199
   undesirable 199
master part file
   See product design using master part file
master part file technique 171
master part technique 201
   product design modeling techniques 79
matching design and plastic technology 51–53
material and mass properties, applying 277–285
material files 277, 280

material flow
  turbulent 153
  See also modeling techniques that enhance material flow
material flow rates
  design planning 58
  maximum diameter reference to gate location 59
material mass properties, applying 277–285
material property and design considerations, other 45–51
material specifications 58
material, flexible 42–45
material, local thinning and thickening of 131–133
materials 51
  hard 51
  strong 51
mating parts, tooling created for all 93
Max Dihedral function for edges and curves 290
Measure Thickness tool 254
memory and plastics properties 44
Merge command 200
Merge Cut Out tool 306
metal
  designer considerations over complex parts, versatility 42
  lack of material properties needed flexible material 44
  prototyping parts made of 30
milling machines, EDM process 14
Minimum Radius tool 291
mirror image part creation 265–269
mirroring project 265–269
mirroring, preparing initial part for 266
misalignments and product design modeling techniques 73
model exterior shape 61
model plan, creation of 100
model planning and root features 87
model top and bottom housing common geometry 180
modeling plan 173–202, 206
modeling plan and process 239–248
modeling process 207
modeling steel safe snap fit joints 106
modeling techniques that enhance material flow 151–164
Modify Component command 268
mold creation, reminders about 133
mold flow analysis for injection molding 23
mold release, rapid tooling materials 32
mold volume, physical device 4
molding machines, efficient values for 19
molding techniques, special
  See thin wall component design
molding, blow, final wall thickness not constant 257
molding, injection, extruded shapes made from die 251
molds, types of 4
  See also blow molding
multi-cavity plastic molding 20

N

neck section, bottle 258
neck, options for creating 258
nest fixture for ultrasonic welding of parts 102
nest 222
  See also assembly jig; nest fixture for ultrasonic welding of parts
no split feature 121
numerically based information tools 288–307

O

offset sketched entity 189
oil canning 181
O-ring groove, parting surfaces and parting planes 86

P

packing pressure of plastic for injection molding, cost considerations of 21
paper, prototyping parts made of 30
parametric model, controlling 162
parent/child relationships
  imported geometry, changes in 321
  root features 87, 88
part
  checking 246
  inside shape of core 5
  outside shape of cavity 5
part creation
  See mirror image part creation
part design of complex parts, versatility 41–42
part design, functioning 49
PART menu 277
part size, increasing
  plastic molding 18
part strength
  See component shape and overall part strength

Index

parting direction 170
parting direction plane, parting surfaces and parting planes 84
parting lines 6
   blow molding 29
   core and cavity separator 6
parting surface, complex curvature
   root features 90
parting surfaces and parting planes 84–92
   O-ring groove 86
parts, tapered 121
   See also stand-alone parts
Pattern option 210
pellets, fabrication methods using 16
per-part cost for plastic molding 20
piece parts
   manufacturability of 70
   outer shape, splitting 66
   See also design splitting; master part technique
piece parts, individual, initial component layout assessment 63
Piezo-electric devices and pressure measurements 19
planes, parting
   See parting surfaces and parting planes
plastic degradation, temperature effects on 17
plastic design, features aiding 123–130
plastic extrusions, post processing on 254
plastic molding
   gates 12
   per-part cost 20
   runners 12
   sprues 12
   types of, multi-cavity 20
plastic molding and fabrication 3–33
plastic molds and rapid prototyping 32
plastic part fabrication methods 15–29
plastic tooling descriptions and terminology 3–13
plastic, cooling and solidification of 152
plastic, mold types 4
plastics
   advantages of use 35
   cost considerations 35
   language of 3
   nonlinear stress/strain curves 43
   other physical properties 53
   preferability of flexible material 43
   properties of 43
   prototyping parts made of 30
   replacement of metal components with 203, 204
   selection of 35–53
   urethane casting 32
plastics design methodology, emphasis placed on 55
plastics design features, shells and features aiding plastics design 115–134
plastics for support structures 253
plastics technology
   See matching design and plastics technology
point cloud data, imported model geometry 311
Poisson's ratio 278
polycarbonate plastic 52
polymers
   air-drying machines 17
   virgin, fabrication methods of 16
Porcupine Analysis tool 296
porcupine curvature tools 294
portable compact disc player, injection molding product design exercise 169–202
pounds per square inch (psi) pressure 19
pressure measurements and Piezo-electric devices 19
proceed feature 319
product design, end user considerations of 56
product design departments, considerations of end users 56
product design modeling techniques 73–80
   datum curves 74
   datum points 75
   master part technique 79
   surfaces 74
product design using master part file 77
product design, component identification 67–73
product designs, visualization of 59–67
   clean sheet approach 59
   design criteria prior to design modeling approach 59
product designs, visualizing
   building-block design 67
   video game controller 67
product external geometry, creating 176
product individual piece part creation 76
product target weight requirements, considerations of end users 57
product's outer shape, laying out 66
product-modeling project, applying design philosophy for 171
project description 170, 204
project planning
   See design philosophy
project, setting up 316

prototyping
   See rapid prototyping
protrusion features, securing interior components and minimizing movement 142
protrusions
   critical height 108
   extrusion from both sides of parting datum plane 181
   solid neck section of bottle 259
   See also Advanced Variable Section Sweep protrusion

Q

Query Select command and root features 88
quick impact 141
quilt option 275
quilt, as joined surface 119

R

radii of gyration 282
Radius Dome feature 130, 140
raised areas and features difficult to import 313
rapid prototyping
   epoxy molds 31
   plastic molds 32
   rapid tooling 31
   undercut parts 31
rapid prototyping and tooling methods 30–33
rapid prototyping parts, x-y plotting methods 30
rapid tooling, rapid prototyping and 31
rapid tooling materials, mold release 32
read-only feature, modeling techniques for 162
recesses, slight, and features difficult to import 313
recycling requirements, considerations of end users 57
recycling, materials and ultrasonic welding of parts 103
REDEFINE menu 323
reference geometry 74
Regen Info option, logical feature creation process and 95
re-import example 322
rename features 89
Replace feature 133
restrictions, identifiable modeling 60
retreat feature 319
reuse of molds 4
revisions, component 94
revolved cut for screw sketch 213

rework and delay reductions, initial design planning for 58
rheology
   See material flow rates
rib and web design techniques and rules 143–149
ribs
   adding internal 245
   alignment 238
   applications, other 149
   crush ribs 219
   curved 148
   deflection resistance 48
   and draft 146
   and draft, curved 148
   guaranteeing space clearance between components 204
   hold-down 178, 192–194, 199, 215
   interlocking 149, 244
   internal, logical feature creation process 97
   rotational 144
   seal-off 204
   straight 143
   techniques and rules 145
   and thin walled designs 148
right-hand model, creating 268
Rockwell scale of hardness 280
root features 176, 177
   buried features 99
   datum curves 89
   design planning and organization 83
   identification of 87–92
   model planning 87
   model with complex surface and datum curves 91
   parent/child relationships 87, 88
   parting surface, complex curvature 90
   Query Select command 88
   recognition of 88
   start part model for product design 174
   tracking features 87
   tracking order 87
   using points as tags 89
roughly modeled subassembly
   known components, gathering and assembling 66
rough-modeling approach, initial component layout assessment 62
rounds
   advanced 129
   automatic blending of 127
   edge chain 124

Index

edge pair 126
edge tangent chain 125
edge-surf 126
energy absorption 47
features difficult to import 313
finishing rounds, adding 230
material flow 154
product design additions 183
simple 124
simple variable radius 128
surf-surf 127
runners
  cost considerations 20
  plastic molding 12
  See also sprues and gates

S

safety
  flame-retardant tests 58
  importance to end users 57
  material specifications 58
sales volume, increased, and considerations of end users 57
Save As option 187
scratch resistant plastic, defined 51
screw boss mounting, modeling of 212
screw bosses
  adding 195, 196, 245
  buried features 98, 99
  clamp inner chassis and secure interlock 193
  cross section 195
  design 111
  dimensioning specifications, identification of critical 94
  internal ribs, adding 245
  logical feature creation process 97
  placement of 236
  project description 204
  seat 235
  start model 179
  use of 109, 110
  wall thickness of 162
screws 234
  shapes and sizes 17
scuff resistance, durable surface finish of 49
seal-off ribs project description 204
sectional assembly
  screw bosses, use of 110
  exaggerated snap fit joints 111
security, importance to end users 57

Set up option 277
Set up section 272
SETUP menu 160
shading tool 300
shadow groove housings, creating top and bottom 184
shape dimensions, modifying, initial component layout assessment for 64
shapes
  basic 260
  creating basic 240
    extruded 251–255
  strengths of 135–150
shear modulus 278
Shell command and plastics design features 115
shell for bottle 263
shells
  constant wall thickness 115
  creating 243
  importance in plastic design 113
  irregular wall thickness 117
  manual creation of 118
  offset inner surfaces of 118
shirt pocket fit requirements, considerations of end users 57
shock guidelines 141
shock or hazardous effects, electrical insulation properties and 36
Short Edge function for edges and curves 290
short shot, injection molding 17
shot size, injection molding 17
shrinkage
  temperature's effect on volume 7
  thin sections versus thick sections 21
shut-off between core and cavity, parting surfaces and parting planes 84
simulated stress tests 285
sink marks 40
  injection molding 21
  managing 156
  material flow 156
skeleton modeling 74, 201
skeleton models 63
  curling iron 74
  visual aids 74
slides
  snap fit feature 10
  undercuts for final plastic part 11
Slope Analysis tool 295
snap features 45, 52

snap fit design 227–236
snap fit design process 228
snap fit joints 104
   modeling steel safe 106
snap fit levers, project description 204
snap fit tab assembly 106
snap fit testing 105
snap hole, creating 232
snap on top housing, locating 228
snap protrusion, adding 229
snap, finished 233
snaps, logical feature creation process for 97
solid geometry and imported model geometry 311
solid protrusion feature 275
sound energy, absorption 46
space claim geometry, skeleton model technique 201
specific heat 279
specifications
   identification of critical 92–95
   known design limitations 56
   product 56
split at curve or quilt 121
split draft, box and tube 87
Split Sketch feature 122
split used as sketch 122
sprues
   cost considerations 20
   plastic molding 12
staking pins 219
stand-alone parts and component fastening 101
standard for the exchange of product model data (STEP) 315–327
starches, edible, prototyping parts made of 30
start part, creating 176
steel mold types 4
steel safe, modeling snap fit joints 106
STEP files, exporting 93
STEP geometry, importing 317
STEP translation systems 309, 310
   imported geometry, redefining tools used for 314
storage size, molded plastic space 17
stress concentration
   energy absorption 47
   material flow 155
   rounds 123
   See also shrinkage
stress limit for compression 279
stress limit for shear 279
stress limit for tension 279
strong plastic, defined 51

structural damping coefficient 279
structural rigidity, ribs and added strength 48
styling features, external 180
support structures
   See plastics for support structures
SURFACE ANALYSIS menu 299
Surface Analysis section 291, 295
surface data, imported model geometry 311
surface models, watertight
   shells of constant wall thickness 116
Surface Normal Vector tool 299
surface stresses, modeling techniques that enhance material flow and 152
surface, replacing of 133
surfaces
   cosmetically acceptable as exterior project description 204
   product design modeling techniques 74
   textured for functional or cosmetic appeal 8
surfacing, neck section of bottle 259
Surf-Surf feature 127
switch positions, component design 172

T

tab ear 132
Tangent Chain option 125
tangent edge chain 186
taper 120
technology limitations, initial design planning and 58
temperature effects and plastic strength 136
temperature measurements, thermocouples 19
temperature, affects on plastic 52
Term Surfs option 127
terminology
   See plastic tooling, descriptions and terminology
test tube area for plastic molding 18
texture requirements 94
textured surfaces,
   cosmetic appearance of decorations 40
thermal conductivity 136, 279
thermal expansion coefficient 278
thermal expansion reference temperature 279
thermal wall designs and strength of shapes 137
thermocouples and temperature measurements 19
thermoplastics 52
Thickness tool 291
thin wall component design, special molding technique 237–249
thin wall design, description of 237

Index

Through/Axis datum plane 144
tiny edge geometry check errors, product design modeling techniques 73
toggling, defined 259
tolerance analysis of complex parts, versatility 42
tolerance modeling, maximum and minimum 160
tonnage
   See pounds per square inch (psi) pressure
tooling methods
   See rapid prototyping
tooling, high expense of for snap fit joints 106
tools, numerically defined
   model categories 287
top and bottom housings, creating 183
top housing door and support features, creating 189
top section, bottle 260
top-down design, visualizing product designs for 59
top-down design approach 60
tough 51
trajectories, neck section of bottle 259
translational acceleration, moment of inertia 281
tube formation in blow molding 27
Tweak command for lip feature 185
Tweak Lip command 186
TWEAK menu 133
tweak offset feature 131
   cosmetic features 115
Tweak Offset function 261

U

ultrasonic vibration 222
ultrasonic welding 102, 221–226
ultrasonic welding process 221
undercut parts in rapid prototyping 31
undercuts, blow molding features 257
urethane mold types 4
urethane casting
   See plastics
using points as tags for root features 89

V

variable ear 131
variable section sweep, door finishing features 190
vents
   EDM process 15
   viscous material tight versus airtight molds 15

vertical EDM process 14
vibration process, ultrasonic welding 102
vibration, ultrasonic 222
Vickers scale of hardness 280
viscous material tight versus airtight molds, vents for 15
viscous plastic
   extrusion molding 25
   foamed parts 23
   mold types 4
   sprues, gates, and runners 13
visual analysis tools 294
visualizing product designs
   bottom-up design approach 64
   exterior shape 61
   layout assessment, initial component 62
volts of discharge, electrical insulation properties 38
volume calculations 271–276
   uses for 271
volume of plastic in a part 272
volume purchasing, bottom-up design approach 64
volume within a plastic part 272
volume, bottle 263

W

wall mount holding tabs 112
warpage, avoiding 252
watertight surface models, shells of constant wall thickness and 116
web design
   See rib and web design techniques and rules
weld lines
   managing 158
   and material flow 156
welding
   See ultrasonic welding
wire-cut EDM process 14
witness line
   See edge mark

X

x-y plotting methods and rapid prototyping parts 30

Y

Young's modulus 278